METHODS IN MOLECULAR BIOLOGY

Series Editor
John M. Walker
School of Life and Medical Sciences
University of Hertfordshire
Hatfield, Hertfordshire, AL10 9AB, UK

For further volumes:
http://www.springer.com/series/7651

Vascular Effects of Hydrogen Sulfide

Methods and Protocols

Edited by

Jerzy Bełtowski

Department of Pathophysiology, Medical University, Lublin, Poland

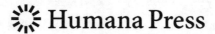 Humana Press

Editor
Jerzy Bełtowski
Department of Pathophysiology
Medical University
Lublin, Poland

ISSN 1064-3745   ISSN 1940-6029   (electronic)
Methods in Molecular Biology
ISBN 978-1-4939-9530-1   ISBN 978-1-4939-9528-8   (eBook)
https://doi.org/10.1007/978-1-4939-9528-8

Preface

Hydrogen sulfide (H_2S) was first identified as the endogenous mediator ("gasotransmitter") in 1996. Since that time, a lot of studies about the role of H_2S in the regulation of various physiological processes have been performed. The vascular system is one of the main H_2S targets. H_2S is synthesized in all parts of the vascular wall (endothelium, smooth muscle cells, periadventitial adipose tissue) and is involved in the regulation of vascular tone, although its exact effect (vasoconstriction vs. vasorelaxation) and mechanism of activity differ depending on experimental animal species and vascular preparation. Alterations of vascular H_2S generation/signaling may be involved in the pathogenesis of systemic and pulmonary arterial hypertension, ischemic heart disease, ischemic stroke, preeclampsia, and erectile dysfunction; all these conditions being associated with abnormal regulation of vascular tone. In addition, H_2S regulates many processes relevant for atherosclerosis such as endothelial activation, inflammatory reaction, uptake of oxidized low density lipoproteins by monocytes/macrophages, and platelet activity. H_2S emerges also as an attractive target for pharmacotherapy of cardiovascular diseases. The role of H_2S in the regulation of angiogenesis and endothelial barrier permeability extends far beyond the cardiovascular system since it is associated with pathologies such as cancer, wound healing, and diabetic retinopathy among others. In this book, some experimental protocols essential for vascular H_2S research are presented by the leading scientists in the field. I hope they will be useful for the researchers interested in this area.

Lublin, Poland *Jerzy Bełtowski*

Contents

Preface . . . . *v*
Contributors . . . . *ix*

1   Synthesis, Metabolism, and Signaling Mechanisms of Hydrogen Sulfide:
An Overview 1
Jerzy Bełtowski

2   Lanthionine and Other Relevant Sulfur Amino Acid Metabolites:
Detection of Prospective Uremic Toxins in Serum by Multiple
Reaction Monitoring Tandem Mass Spectrometry 9
Alessandra F. Perna, Francesca Pane, Nunzio Sepe, Carolina Fontanarosa,
Gabriella Pinto, Miriam Zacchia, Francesco Trepiccione, Evgeniya
Anishchenko, Diego Ingrosso, Piero Pucci, and Angela Amoresano

3   Analysis of Vascular Hydrogen Sulfide Biosynthesis. 19
Thomas J. Lechuga and Dong-bao Chen

4   Measurement of Protein Persulfidation: Improved Tag-Switch Method 37
Emilia Kouroussis, Bikash Adhikari, Jasmina Zivanovic,
and Milos R. Filipovic

5   ProPerDP: A Protein Persulfide Detection Protocol 51
Éva Dóka, Elias S. J. Arnér, Edward E. Schmidt, and Péter Nagy

6   Vascular Effects of H_2S-Donors: Fluorimetric Detection of H_2S
Generation and Ion Channel Activation in Human Aortic
Smooth Muscle Cells 79
Alma Martelli, Valentina Citi, and Vincenzo Calderone

7   In Vitro Measurement of H_2S-Mediated Vasoactive Responses 89
Sona Cacanyiova and Andrea Berenyiova

8   In Vivo Measurement of H_2S, Polysulfides, and "SSNO⁻
Mix"-Mediated Vasoactive Responses and Evaluation of Ten
Hemodynamic Parameters from Rat Arterial Pulse Waveform 109
Frantisek Kristek, Marian Grman, and Karol Ondrias

9   Simultaneous Measurements of Tension and Free H_2S
in Mesenteric Arteries 125
Elise Røge Nielsen, Anna K. Winther, and Ulf Simonsen

10 The Relaxant Mechanisms of Hydrogen Sulfide in Corpus
Cavernosum. 137
Fatma Aydinoglu and Nuran Ogulener

11 Pharmacological Tools for the Study of H_2S Contribution
to Angiogenesis. 151
Lucia Morbidelli, Martina Monti, and Erika Terzuoli

12 Central Administration of H_2S Donors for Studying
Cardiovascular Effects of H_2S in Rats 167
Marcin Ufnal and Artur Nowinski

13   Colonic Delivery of H_2S Donors for Studying Cardiovascular Effects
     of H_2S in Rats.. 173
     Marcin Ufnal and Tomasz Hutsch

14   Measurements for Sulfide-Mediated Inhibition of Myeloperoxidase
     Activity.. 179
     *Dorottya Garai, Zoltán Pálinkás, József Balla, Anthony J. Kettle,
     and Péter Nagy*

15   Vascular Myography to Examine Functional Responses of Isolated
     Blood Vessels.. 205
     Joanne Hart

Index *219*

Contributors

BIKASH ADHIKARI • *Université de Bordeaux, IBGC, CNRS UMR 5095, Bordeaux, France; Institut de Biochimie et Génétique Cellulaires, CNRS, UMR 5095, Bordeaux, France*

ANGELA AMORESANO • *Department of Chemical Sciences, University of Naples "Federico II", Naples, Italy*

EVGENIYA ANISHCHENKO • *First Division of Nephrology, Department of Translational Medical Sciences, School of Medicine, University of Campania "Luigi Vanvitelli", Naples, Italy*

ELIAS S. J. ARNÉR • *Division of Biochemistry, Department of Medical Biochemistry and Biophysics, Karolinska Institutet, Stockholm, Sweden*

FATMA AYDINOGLU • *Department of Pharmacology, Pharmacy Faculty, Cukurova University, Adana, Turkey*

JÓZSEF BALLA • *HAS-UD Vascular Biology and Myocardial Pathophysiology Research Group, Hungarian Academy of Sciences, Debrecen, Hungary*

JERZY BEŁTOWSKI • *Department of Pathophysiology, Medical University, Lublin, Poland*

ANDREA BERENYIOVA • *Institute of Normal and Pathological Physiology, Center of Experimental Medicine, Slovak Academy of Sciences, Bratislava, Slovakia*

SONA CACANYIOVA • *Institute of Normal and Pathological Physiology, Center of Experimental Medicine, Slovak Academy of Sciences, Bratislava, Slovakia*

VINCENZO CALDERONE • *Department of Pharmacy, University of Pisa, Pisa, Italy*

DONG-BAO CHEN • *Department of Obstetrics and Gynecology & Pathology, University of California Irvine, Irvine, CA, USA*

VALENTINA CITI • *Department of Pharmacy, University of Pisa, Pisa, Italy*

ÉVA DÓKA • *Department of Molecular Immunology and Toxicology, National Institute of Oncology, Budapest, Hungary*

MILOS R. FILIPOVIC • *Université de Bordeaux, IBGC, CNRS UMR 5095, Bordeaux, France; Institut de Biochimie et Génétique Cellulaires, CNRS, UMR 5095, Bordeaux, France*

CAROLINA FONTANAROSA • *Department of Chemical Sciences, University of Naples "Federico II", Naples, Italy*

DOROTTYA GARAI • *Department of Molecular Immunology and Toxicology, National Institute of Oncology, Budapest, Hungary; Faculty of Medicine, Laki Kálmán Doctoral School, University of Debrecen, Debrecen, Hungary*

MARIAN GRMAN • *Institute of Clinical and Translational Research, Biomedical Research Center, Slovak Academy of Sciences, Bratislava, Slovakia*

JOANNE HART • *Faculty of Medicine and Health, School of Medicine, University of Sydney, Camperdown, NSW, Australia*

TOMASZ HUTSCH • *Laboratory of Centre for Preclinical Research, Department of Experimental Physiology and Pathophysiology, Medical University of Warsaw, Warsaw, Poland*

DIEGO INGROSSO • *Department of Precision Medicine, School of Medicine, University of Campania "Luigi Vanvitelli", Naples, Italy*

ANTHONY J. KETTLE • *Centre for Free Radical Research Department of Pathology and Biomedical Science, University of Otago Christchurch, Christchurch, New Zealand*

EMILIA KOUROUSSIS • *Université de Bordeaux, IBGC, CNRS UMR 5095, Bordeaux, France; Institut de Biochimie et Génétique Cellulaires, CNRS, UMR 5095, Bordeaux, France*

FRANTISEK KRISTEK • *Institute of Normal and Pathological Physiology, Centre of Experimental Medicine, Slovak Academy of Sciences, Bratislava, Slovakia*

THOMAS J. LECHUGA • *Department of Obstetrics and Gynecology & Pathology, University of California Irvine, Irvine, CA, USA*

ALMA MARTELLI • *Department of Pharmacy, University of Pisa, Pisa, Italy*

MARTINA MONTI • *Department of Life Sciences, University of Siena, Siena, Italy*

LUCIA MORBIDELLI • *Department of Life Sciences, University of Siena, Siena, Italy*

PÉTER NAGY • *Department of Molecular Immunology and Toxicology, National Institute of Oncology, Budapest, Hungary; Department of Medicine, Faculty of Medicine, University of Debrecen, Debrecen, Hungary*

ELISE RØGE NIELSEN • *Department of Biomedicine, Pharmacology, Aarhus University, Aarhus C, Denmark*

ARTUR NOWINSKI • *Laboratory of Centre for Preclinical Research, Department of Experimental Physiology and Pathophysiology, Medical University of Warsaw, Warsaw, Poland*

NURAN OGULENER • *Department of Pharmacology, Medical Faculty, Cukurova University, Adana, Turkey*

KAROL ONDRIAS • *Institute of Clinical and Translational Research, Biomedical Research Center, Slovak Academy of Sciences, Bratislava, Slovakia*

ZOLTÁN PÁLINKÁS • *Department of Molecular Immunology and Toxicology, National Institute of Oncology, Budapest, Hungary*

FRANCESCA PANE • *Department of Chemical Sciences, University of Naples "Federico II", Naples, Italy*

ALESSANDRA F. PERNA • *First Division of Nephrology, Department of Translational Medical Sciences, School of Medicine, University of Campania "Luigi Vanvitelli", Naples, Italy*

GABRIELLA PINTO • *Department of Chemical Sciences, University of Naples "Federico II", Naples, Italy*

PIERO PUCCI • *Department of Chemical Sciences, University of Naples "Federico II", Naples, Italy*

EDWARD E. SCHMIDT • *Department of Microbiology and Immunology, Montana State University, Bozeman, MT, USA*

NUNZIO SEPE • *Department of Chemical Sciences, University of Naples "Federico II", Naples, Italy*

ULF SIMONSEN • *Department of Biomedicine, Pharmacology, Aarhus University, Aarhus C, Denmark*

ERIKA TERZUOLI • *Department of Life Sciences, University of Siena, Siena, Italy*

FRANCESCO TREPICCIONE • *First Division of Nephrology, Department of Translational Medical Sciences, School of Medicine, University of Campania "Luigi Vanvitelli", Naples, Italy*

MARCIN UFNAL • *Laboratory of Centre for Preclinical Research, Department of Experimental Physiology and Pathophysiology, Medical University of Warsaw, Warsaw, Poland*

ANNA K. WINTHER • *Department of Chemistry, Aarhus University, Aarhus C, Denmark*

MIRIAM ZACCHIA • *First Division of Nephrology, Department of Translational Medical Sciences, School of Medicine, University of Campania "Luigi Vanvitelli", Naples, Italy*

JASMINA ZIVANOVIC • *Université de Bordeaux, IBGC, CNRS UMR 5095, Bordeaux, France; Institut de Biochimie et Génétique Cellulaires, CNRS, UMR 5095, Bordeaux, France*

Chapter 1

Synthesis, Metabolism, and Signaling Mechanisms of Hydrogen Sulfide: An Overview

Jerzy Bełtowski

Abstract

In addition to nitric oxide (NO) and carbon monoxide (CO), hydrogen sulfide (H_2S) has recently emerged as the novel gasotransmitter involved in the regulation of the nervous system, cardiovascular functions, inflammatory response, gastrointestinal system, and renal function. H_2S is synthesized from L-cysteine and/or L-homocysteine by cystathionine β-synthase, cystathionine γ-lyase, and cysteine aminotransferase together with 3-mercaptopyruvate sulfurtransferase. In addition, H_2S is enzymatically metabolized in mitochondria by sulfide:quinone oxidoreductase, persulfide dioxygenase, and sulfite oxidase to thiosulfate, sulfite, and sulfate which enables to regulate its level by factors such as oxygen pressure, mitochondria density, or efficacy of mitochondrial electron transport. H_2S modifies protein structure and function through the so-called sulfuration or persulfidation, that is, conversion of cysteine thiol (−SH) to persulfide (−SSH) groups. This, as well as other signaling mechanisms, is partially mediated by more oxidized H_2S-derived species, polysulfides (H_2S_n). In addition, H_2S is able to react with reactive oxygen and nitrogen species to form other signaling molecules such as thionitrous acid (HSNO), nitrosopersulfide ($SSNO^-$), and nitroxyl (HNO). All H_2S-synthesizing enzymes are expressed in the vascular wall, and H_2S has been demonstrated to regulate vascular tone, endothelial barrier permeability, angiogenesis, vascular smooth muscle cell proliferation and apoptosis, and inflammatory reaction. H_2S-modifying therapies are promising approach for diseases such as arterial hypertension, diabetic angiopathy, and atherosclerosis.

Key words Hydrogen sulfide, Gasotransmitters, Polysulfides, Vascular tone, Arterial hypertension, Atherosclerosis

1 Introduction

H_2S has been known for a long time only as the environmental toxin until Abe and Kimura first proposed its role as the endogenously produced signaling molecule [1]. Currently, it is quite apparent that hydrogen sulfide is the third "gasotransmitter" in addition to nitric oxide (NO) and carbon monoxide (CO) [2]. H_2S is involved in the regulation of many biological processes such as neurotransmission, inflammatory and immune response, gastrointestinal system function, and vascular homeostasis. In the vascular system, H_2S is involved in the regulation of

Jerzy Bełtowski (ed.), *Vascular Effects of Hydrogen Sulfide: Methods and Protocols*, Methods in Molecular Biology, vol. 2007, https://doi.org/10.1007/978-1-4939-9528-8_1, © Springer Science+Business Media, LLC, part of Springer Nature 2019

vascular tone, angiogenesis, smooth muscle cell growth and apoptosis, as well as lipid accumulation in the atherosclerotic plaque making it a potential target for therapy of some civilization diseases such as arterial hypertension or atherosclerosis [3]. Vascular effects of H_2S are being intensively examined in multiple laboratories. In the subsequent chapters are presented selected experimental protocols relevant for vascular H_2S research. In this chapter, I present some general aspects of H_2S biology to provide the reader the brief introduction to the topic.

2   General Properties of H_2S

H_2S is a colorless flammable gas with a strong odor of rotten eggs. It is easily soluble in both water and lipids and as such easily crosses plasma membranes [3]. At physiological conditions, about 20% of H_2S exists in the undissociated form, and the rest is dissociated to HS^- (hydrosulfide anion) and H^+, the amount of sulfide anion (S^{2-}) being very low at physiological pH. Both H_2S and HS^- always coexist in solution, and it is not possible to separate their effects and to conclude which of them is involved in signaling processes [4]. The pK_a of H_2S is 6.8 that is 1–2 units lower than of other thiols such as glutathione (GSH) or protein cysteine thiol groups. Therefore, the greater portion of H_2S exists in the dissociated form (HS^-) in comparison to other thiols [4].

3   Endogenous H_2S Synthesis

Endogenous H_2S is synthesized from L-cysteine by cystathionine β-synthase (CBS) and cystathionine γ-lyase (CSE), two pyridoxal 5′-phosphate (vitamin B_6)-dependent enzymes [5]. CBS and CSE are involved in transsulfuration pathway of homocysteine metabolism. CBS condenses L-homocysteine and L-serine to form L-cystathionine and H_2O. L-serine can be replaced by L-cysteine in this reaction resulting in H_2S formation. The activity of CBS toward L-cysteine is much lower than toward L-serine; therefore, the rate of H_2S synthesis is much lower than the rate of "canonical" reaction in the transsulfuration pathway. It should be noted that H_2S synthesis by CBS requires simultaneous presence of L-cysteine and L-homocysteine [6]. CBS may also catalyze the alternative reactions between two cysteine or two homocysteine molecules to form H_2S and lanthionine (two cysteine molecules connected by thioeter –S- bond) or homolanthionine (two homocysteine molecules connected by the –S- bond), respectively [6]. S-adenosylmethionine is the allosteric CBS activator, whereas two remaining gasotransmitters, NO and CO, inhibit the enzyme by binding to its heme group [7, 8].

In the transsulfuration pathway, CSE metabolizes L-cystathionine to L-cysteine, α-ketobutyrate, and ammonia. The main reaction catalyzed by CSE and leading to H_2S formation is the β-elimination of L-cysteine or γ-elimination of L-homocysteine to ammonia, H_2S, and pyruvate or α-ketobutyrate, respectively [5]. Because L-cysteine concentration exceeds that of L-homocysteine, cysteine desulfhydration is the main mechanism of H_2S production by CSE under physiological conditions [9]. Hyperhomocysteinemia is associated with the increase in H_2S synthesis by both CBS and CSE. Both CBS and CSE can break down L-cystine (cysteine disulfide) to thiocysteine (CysSSH, cysteine persulfide), pyruvate, and ammonia [10]. Through further exchange reactions between CysSSH and glutathione or protein cysteine thiols, the respective persulfide species (GSSH and protein-CysSSH) may then be formed.

The third mechanism of H_2S generation is associated with conversion of L-cysteine to 3-mercaptopytuvate (3-MP) by cysteine aminotransferase (CAT, aspartate aminotransferase); 3-MPis the substrate for 3-mercaptopyruvate sulfurtransferase (3-MST) which, in the presence of dithiol-containing reductants such as thioredoxin or dihydrolipoic acid, converts 3-mercaptopyruvate to pyruvate and H_2S [11, 12]. Thioredoxin and dihydrolipoic acid increase 3-MST-catalyzed H_2S formation by ten- and threefold, respectively, while other reductants such as NADH, NADPH, or coenzyme A are without effect. It is believed that 3-MST transfers sulfur atom from 3-MP to the enzyme itself forming 3-MST persulfide from which H_2S is subsequently released by the reducing agent. H_2S synthesis by CAT/3-MST pathway is maximal in the absence of calcium and is markedly inhibited by the increase in intracellular calcium concentration [11–13].

H_2S may also be synthesized from D-cysteine by sequential activities of D-amino acid oxidase (DAO) and 3-MST. This pathway has been so far demonstrated only in the kidney and cerebellum because DAO is abundantly expressed in peroxisomes of these organs. D-cysteine is not the endogenous amino acid in mammals but may be present in food as well as is synthesized by gut bacteria by racemization of L-amino acid. After administration of exogenous D-cysteine, the rate of H_2S production by the kidney is much higher than in the presence of L-cysteine, and as such D-cysteine offers the possibility of more potent and more organ-specific tissue protection [14].

4   H₂S Metabolism

Interestingly, H_2S is the only known inorganic substrate which is enzymatically oxidized in mitochondria providing the evidence for the origin of mitochondria from sulfide-oxidizing bacteria. H_2S is oxidized by sulfide:quinone oxidoreductase (SQR) [15], which transfers electrons to ubiquinone (coenzyme Q) from which they are further transported to cytochrome c by ubiquinol:cytochrome c oxidase (complex III) and finally to molecular oxygen by cytochrome c oxidase (complex IV) [16, 17]. SQR represents the third source of electrons for mitochondrial respiratory chain in addition to NADH dehydrogenase (complex I) and succinate dehydrogenase (complex II) involved in oxidation of organic substrates. During SQR-catalyzed reaction, H_2S is oxidized to sulfur atom incorporated into SQR molecule as cysteine persulfide. This sulfur atom is then transferred to reduced glutathione to form glutathione persulfide (GSSH) and then may be either oxidized to sulfite (SO_3^{2-}) by sulfur dioxygenase (persulfide dioxygenase, protein deficient in ethylmalonic encephalopathy, ETHE1) or transferred to sulfite by thiosulfate:cyanide sulfurtransferase (TST; rhodanese) to form thiosulfate (SSO_3^{2-}) [18, 19]. Sulfite is further oxidized to sulfate (SO_4^{2-}) by sulfite oxidase. It should be noted that H_2S is oxidized in mitochondria only at very low concentrations; at concentrations higher than 5 μM, it potently inhibits cytochrome c oxidase, thus leading to metabolic suppression and ATP deficiency.

The rate of mitochondrial H_2S oxidation is determined by oxygen level. The results of many studies strongly suggest that steady-state level of H_2S in tissues is maintained very low by its efficient oxidation. However, under conditions of reduced O_2 supply, H_2S metabolism is compromised, and its concentration increases making this gasotransmitter an important oxygen sensor in various tissues including blood vessels. Indeed, H_2S may at least partially mediate some effects of hypoxia [20, 21].

5   H$_2$S Signaling

According to current knowledge, H_2S interacts with three kinds of target molecules, protein or low-molecular thiol groups, reactive oxygen and nitrogen species, and metalloproteins.

H_2S interacts with protein cysteine thiol(–SH) groups converting them to hydrosulfide (–SSH), the process referred to as sulfuration or persulfidation [22]. The examples of protein modified in this way include ATP-sensitive potassium channels (K_{ATP}), small and intermediate conductance calcium-activated potassium channels (SK_{Ca}, IK_{Ca} channels), voltage-sensitive K^+ channel Kv4.3, vascular endothelial growth factor receptor-1 (VEGFR1), ubiquitin E3 ligase parkin, p65 subunit of nuclear factor-κB (NF-κB), Kelch-like ECH-associated protein 1 (Keap1), endothelial NO synthase (eNOS), protein kinase G (PKG), protein tyrosine phosphatase 1B (PTP1B), phospholamban, MAPK/ERK kinase (MEK1), transcriptional repressor interferon regulatory factor 1 (IRF1), and TRPV6 calcium channel and glyceraldehyde 3-phosphate dehydrogenase (GAPDH) [23–33]. Both stimulatory and inhibitory effects of H_2S on protein activity have been described. The persulfide group has usually lower pK_a value that is more acidic and more easily dissociates to perthiolate anion (R-SS$^-$), which is a better reductant and more electrophilic than the respective thiol [34].

H_2S cannot directly sulfurate proteins; the reaction requires prior oxidation of thiol group to disulfides or sulfenic acid (–SOH) or oxidation of H_2S itself to inorganic polysulfides (H_2S_n). Indeed, polysulfides are detected in tissues and in some experimental systems are much more effective than H_2S itself [35, 36].

H_2S has been demonstrated to interact with and to reduce various reactive oxygen species such as superoxide anion radical (O_2^-), hydrogen peroxide (H_2O_2), peroxynitrite (ONOO$^-$), and hypochlorous acid (HOCl). However, it is unclear if these reactions are relevant in vivo due to very low H_2S concentration. Rather, H_2S may protect tissues from oxidative stress by indirect mechanisms including stimulation of glutathione synthesis and activation of Nrf2 transcription factor-driven expression of antioxidant proteins such as thioredoxin or heme oxygenase-1 (HO-1).

In addition, H_2S interacts with NO to form thionitrous acid (HSNO) resulting in NO scavenging and vasoconstriction at low H_2S concentration [37]. H_2S may also react with protein or low molecular weight nitrosothiols such as S-nitroso-glutathione (GSNO) and restore thiol group or to release free NO thus potentiating its effects on soluble guanylate cyclase. Another product of the reaction between nitrosothiols and H_2S is nitrosopersulfide (SSNO$^-$), a stable activator of soluble guanylate cyclase. SSNO$^-$ decomposes to nitrosonium cation (NO$^+$) and disulfide (S_2^{2-}); in this way NO catalyzes polysulfide formation from H_2S [38–40]. H_2S may also react with HSNO leading to the formation of hydrogen disulfide (the simplest polysulfide) and nitroxyl (HNO) [41]. HNO, similarly to NO, induces vasorelaxation by activating soluble guanylyl cyclase as well as improves systolic and diastolic function of the heart [42]. The other mechanism of HNO-induced vasorelaxation is activation of transient receptor potential ankyrin 1 (TRPA1) channels in perivascular sensory nerves and stimulating release of calcitonin gene-related peptide (CGRP) [43].

Finally, H_2S interacts with some hemeproteins such as cytochrome c oxidase, hemoglobin, myoglobin, and myeloperoxidase [44]. Moderate inhibition of mitochondrial respiration may be protective under hypoxic conditions such as ischemia-reperfusion injury. H_2S inhibits neutrophil myeloperoxidase (MPO) which catalyzes the reaction between H_2O_2 and chloride (Cl$^-$) to form hypochlorous acid (HOCl). This mechanism may play an important role in anti-inflammatory effect of H_2S [45].

All H_2S-synthesizing enzymes are expressed in the vascular wall including endothelial cells, smooth muscle cells, and perivascular adipose tissue. H_2S is involved in the regulation of vascular tone, smooth muscle cell proliferation and apoptosis, lipoprotein uptake by macrophages, and local vascular inflammatory reaction. In addition, H_2S may regulate endothelial barrier permeability and angiogenesis. Although a lot of data about vascular effects of this gasotransmitter have been accumulated over time, this will remain the hot area of research in the near future.

References

1. Abe K, Kimura H (1966) The possible role of hydrogen sulfide as an endogenous neuromodulator. J Neurosci 16:1066–1071
2. Wang R (2014) Gasotransmitters: growing pains and joys. Trends Biochem Sci 39:227–232
3. Wang R (2012) Physiological implications of hydrogen sulfide: a whiff exploration that blossomed. Physiol Rev 92:791–896
4. Li Q, Lancaster JR (2013) Chemical foundations of hydrogen sulfide biology. Nitric Oxide 35:21–34
5. Kabil O, Banerjee R (2014) Enzymology of H_2S biogenesis, decay and signaling. Antioxid Redox Signal 20:770–782
6. Singh S, Padovani D, Leslie RA, Chiku T, Banerjee R (2009) Relative contributions of cystathionine β-synthase and γ-cystathionase to H_2S biogenesis via alternative transsulfuration reactions. J Biol Chem 284:22457–22466
7. Vicente JB, Colaço HG, Mendes MI, Sarti P, Leandro P, Giuffrè A (2014) NO* binds human cystathionine β-synthase quickly and tightly. J Biol Chem 289:8579–8587
8. Carballal S, Cuevasanta E, Marmisolle I, Kabil O, Gherasim C, Ballou DP, Banerjee R, Alvarez B (2013) Kinetics of reversible reductive carbonylation of heme in human cystathionine β-synthase. Biochemistry 52:4553–4562
9. Chiku T, Padovani D, Zhu W, Singh S, Vitvitsky V, Banerjee R (2009) H_2S biogenesis by human cystathionine γ-lyase leads to the novel sulfur metabolites lanthionine and homolanthionine and is responsive to the grade of hyperhomocysteinemia. J Biol Chem 284:11601–11612
10. Ida T, Sawa T, Ihara H, Tsuchiya Y, Watanabe Y, Kumagai Y, Suematsu M, Motohashi H, Fujii S, Matsunaga T, Yamamoto M, Ono K, Devarie-Baez NO, Xian M, Fukuto JM, Akaike T (2014) Reactive cysteine persulfides and S-polythiolation regulate oxidative stress and redox signaling. Proc Natl Acad Sci U S A 111:7606–7611
11. Shibuya N, Tanaka M, Yoshida M, Ogasawara Y, Togawa T, Ishii K, Kimura H (2009) 3-Mercaptopyruvate sulfurtransferase produces hydrogen sulfide and bound sulfane sulfur in the brain. Antioxid Redox Signal 11:703–714
12. Shibuya N, Mikami Y, Kimura Y, Nagahara N, Kimura H (2009) Vascular endothelium expresses 3-mercaptopyruvate sulfurtransferase and produces hydrogen sulfide. J Biochem 146:623–626
13. Mikami Y, Shibuya N, Kimura Y, Nagahara N, Ogasawara Y, Kimura H (2011) Thioredoxin and dihydrolipoic acid are required for 3-mercaptopyruvate sulfurtransferase to produce hydrogen sulfide. Biochem J 439:479–485
14. Shibuya N, Koike S, Tanaka M, Ishigami-Yuasa M, Kimura Y, Ogasawara Y, Fukui K, Nagahara N, Kimura H (2013) A novel pathway for the production of hydrogen sulfide from D-cysteine in mammalian cells. Nat Commun 4:1366
15. Hildebrandt TM, Grieshaber MK (2008) Three enzymatic activities catalyze the oxidation of sulfide to thiosulfate in mammalian and invertebrate mitochondria. FEBS J 275:3352–3361
16. Bouillaud F, Blachier F (2011) Mitochondria and sulfide: a very old story of poisoning, feeding, and signaling? Antioxid Redox Signal 15:379–391
17. Jackson MR, Melideo SL, Jorns MS (2012) Human sulfide:quinone oxidoreductase catalyzes the first step in hydrogen sulfide metabolism and produces a sulfane sulfur metabolite. Biochemistry 51:6804–6815
18. Olson KR (2012) Mitochondrial adaptations to utilize hydrogen sulfide for energy and signaling. J Comp Physiol B 182:881–897

19. Libiad M, Yadav PK, Vitvitsky V, Martinov M, Banerjee R (2014) Organization of the human mitochondrial hydrogen sulfide oxidation pathway. J Biol Chem 289:30901–30910

20. Olson KR (2008) Hydrogen sulfide and oxygen sensing: implications in cardiorespiratory control. J Exp Biol 211:2727–2734

21. Olson KR (2011) Hydrogen sulfide is an oxygen sensor in the carotid body. Respir Physiol Neurobiol 179:103–110

22. Mustafa AK, Gadalla MM, Sen N, Kim S, Mu W, Gazi SK, Barrow RK, Yang G, Wang R, Snyder SH (2009) H_2S signals through protein S-sulfhydration. Sci Signal 2: ra72

23. Mustafa AK, Sikka G, Gazi SK, Steppan J, Jung SM, Bhunia AK, Barodka VM, Gazi FK, Barrow RK, Wang R, Amzel LM, Berkowitz DE, Snyder SH (2011) Hydrogen sulfide as endothelium-derived hyperpolarizing factor sulfhydrates potassium channels. Circ Res 109:1259–1268

24. Krishnan N, Fu C, Pappin DJ, Tonks NK (2011) H_2S-Induced sulfhydration of the phosphatase PTP1B and its role in the endoplasmic reticulum stress response. Sci Signal 4: ra86

25. Sen N, Paul BD, Gadalla MM, Mustafa AK, Sen T, Xu R, Kim S, Snyder SH (2012) Hydrogen sulfide-linked sulfhydration of NF-κB mediates its antiapoptotic actions. Mol Cell 45:13–24

26. Yang G, Zhao K, Ju Y, Mani S, Cao Q, Puukila S, Khaper N, Wu L, Wang R (2013) Hydrogen sulfide protects against cellular senescence via S-sulfhydration of Keap1 and activation of Nrf2. Antioxid Redox Signal 18:1906–1919

27. Guo C, Liang F, Shah Masood W, Yan X (2014) Hydrogen sulfide protected gastric epithelial cell from ischemia/reperfusion injury by Keap1 S-sulfhydration, MAPK dependent anti-apoptosis and NF-κB dependent anti-inflammation pathway. Eur J Pharmacol 725:70–78

28. Altaany Z, Ju Y, Yang G, Wang R (2014) The coordination of S-sulfhydration, S-nitrosylation, and phosphorylation of endothelial nitric oxide synthase by hydrogen sulfide. Sci Signal 7:ra87

29. Vandiver MS, Paul BD, Xu R, Karuppagounder S, Rao F, Snowman AM, Ko HS, Lee YI, Dawson VL, Dawson TM, Sen N, Snyder SH (2013) Sulfhydration mediates neuroprotective actions of parkin. Nat Commun 4:1626

30. Mazza R, Pasqua T, Cerra MC, Angelone T, Gattuso A (2013) Akt/eNOS signaling and PLN S-sulfhydration are involved in H_2S-dependent cardiac effects in frog and rat. Am J Physiol Regul Integr Comp Physiol 305: R443–R451

31. Liu DH, Huang X, Meng XM, Zhang CM, Lu HL, Kim YC, Xu WX (2014) Exogenous H_2S enhances mice gastric smooth muscle tension through S-sulfhydration of Kv4.3, mediating the inhibition of the voltage-dependent potassium current. Neurogastroenterol Motil 26:1705–1716

32. Zhao K, Ju Y, Li S, Altaany Z, Wang R, Yang G (2014) S-sulfhydration of MEK1 leads to PARP-1 activation and DNA damage repair. EMBO Rep 15:792–800

33. Stubbert D, Prysyazhna O, Rudyk O, Scotcher J, Burgoyne JR, Eaton P (2014) Protein kinase G Iα oxidation paradoxically underlies blood pressure lowering by the reductant hydrogen sulfide. Hypertension 64:1344–1351

34. Ono K, Akaike T, Sawa T, Kumagai Y, Wink DA, Tantillo DJ, Hobbs AJ, Nagy P, Xian M, Lin J, Fukuto JM (2014) Redox chemistry and chemical biology of H_2S, hydropersulfides, and derived species: implications of their possible biological activity and utility. Free Radic Biol Med 77:82–94

35. Greiner R, Pálinkás Z, Bäsell K, Becher D, Antelmann H, Nagy P, Dick TP (2013) Polysulfides link H_2S to protein thiol oxidation. Antioxid Redox Signal 19:1749–1765

36. Koike S, Ogasawara Y, Shibuya N, Kimura H, Ishii K (2013) Polysulfide exerts a protective effect against cytotoxicity caused by t-buthylhydroperoxide through Nrf2 signaling in neuroblastoma cells. FEBS Lett 587:3548–3555

37. Filipovic MR, JL M, Nauser T, Royzen M, Klos K, Shubina T, Koppenol WH, Lippard SJ, Ivanović-Burmazović I (2012) Chemical characterization of the smallest S-nitrosothiol, HSNO; cellular cross-talk of H_2S and S-nitrosothiols. J Am Chem Soc 134:12016–12027

38. Bruce King S (2013) Potential biological chemistry of hydrogen sulfide (H_2S) with the nitrogen oxides. Free Radic Biol Med 55:1–7

39. Ondrias K, Stasko A, Cacanyiova S, Sulova Z, Krizanova O, Kristek F, Malekova L, Knezl V, Breier A (2008) H_2S and HS^- donor NaHS releases nitric oxide from nitrosothiols, metal nitrosyl complex, brain homogenate and murine L1210 leukaemia cells. Pflugers Arch 457:271–279

40. Cortese-Krott MM, Fernandez BO, Santos JL, Mergia E, Grman M, Nagy P, Kelm M, Butler A, Feelisch M (2014) Nitrosopersulfide (SSNO⁻) accounts for sustained NO bioactivity of S-nitrosothiols following reaction with sulfide. Redox Biol 2:234–244

41. Yong QC, Hu LF, Wang S, Huang D, Bian JS (2010) Hydrogen sulfide interacts with nitric oxide in the heart: possible involvement of nitroxyl. Cardiovasc Res 88:482–491

42. Sivakumaran V, Stanley BA, Tocchetti CG, Ballin JD, Caceres V, Zhou L, Keceli G, Rainer PP, Lee DI, Huke S, Ziolo MT, Kranias EG, Toscano JP, Wilson GM, O'Rourke B, Kass DA, Mahaney JE, Paolocci N (2013) HNO enhances SERCA2a activity and cardiomyocyte function by promoting redox-dependent phospholamban oligomerization. Antioxid Redox Signal 19:1185–1197

43. Eberhardt M, Dux M, Namer B, Miljkovic J, Cordasic N, Will C, Kichko TI, de la Roche J, Fischer M, Suárez SA, Bikiel D, Dorsch K, Leffler A, Babes A, Lampert A, Lennerz JK, Jacobi J, Martí MA, Doctorovich F, Högestätt ED, Zygmunt PM, Ivanovic-Burmazovic I, Messlinger K, Reeh P, Filipovic MR (2014) H₂S and NO cooperatively regulate vascular tone by activating a neuroendocrine HNO-TRPA1-CGRP signalling pathway. Nat Commun 5:4381

44. Pietri R, Román-Morales E, López-Garriga J (2011) Hydrogen sulfide and hemeproteins: knowledge and mysteries. Antioxid Redox Signal 15:393–404

45. Pálinkás Z, Furtmüller PG, Nagy A, Jakopitsch C, Pirker KF, Magierowski M, Jasnos K, Wallace JL, Obinger C, Nagy P (2015) Interactions of hydrogen sulfide with myeloperoxidase. Br J Pharmacol 172:1516–1532

Chapter 2

Lanthionine and Other Relevant Sulfur Amino Acid Metabolites: Detection of Prospective Uremic Toxins in Serum by Multiple Reaction Monitoring Tandem Mass Spectrometry

Alessandra F. Perna, Francesca Pane, Nunzio Sepe, Carolina Fontanarosa, Gabriella Pinto, Miriam Zacchia, Francesco Trepiccione, Evgeniya Anishchenko, Diego Ingrosso, Piero Pucci, and Angela Amoresano

Abstract

In the context of the vascular effects of hydrogen sulfide (H_2S), it is known that this gaseous endogenous biological modulator of inflammation, oxidative stress, etc. is a potent vasodilator. Chronic renal failure, a common disease affecting the aging population, is characterized by low levels of H_2S in plasma and tissues, which could mediate their typical hypertensive pattern, along with other abnormalities. Lanthionine and homolanthionine, natural non-proteinogenic amino acids, are formed as side products of H_2S production. Also in consideration of the intrinsic difficulties in H_2S measuring, these compounds have been proposed as reliable and stable markers of H_2S synthesis. However, in the setting of chronic renal failure patients on hemodialysis, they represent typical retention products (without ruling out the possibility of an increased intestinal synthesis) and prospective novel uremic toxins. Here, a method utilizing liquid chromatography-electrospray tandem mass spectrometry (LC-MS/MS) in multiple reaction monitoring ion mode has been developed and evaluated for the determination of these key H_2S metabolites in plasma, by using a triple quadrupole mass spectrometer.

Key words Mass spectrometry, Multiple reaction monitoring, Targeted analysis, Metabolomics, Homocysteine, Homoserine, Lanthionine, Cystathionine

1 Introduction

Hydrogen sulfide (H_2S) is an important gasotransmitter modulating several biological functions, ranging from lifespan extension to regulation of vascular tone, anti-oxidative and anti-inflammatory effects, etc. Low plasma and tissue H_2S have been detected in chronic renal disease (CKD), which has been attributed to a down-regulation of cystathionine gamma lyase (CSE), one of the main

Jerzy Bełtowski (ed.), *Vascular Effects of Hydrogen Sulfide: Methods and Protocols*, Methods in Molecular Biology, vol. 2007,
https://doi.org/10.1007/978-1-4939-9528-8_2, © Springer Science+Business Media, LLC, part of Springer Nature 2019

H$_2$S-forming enzymes. CKD is a common occurrence worldwide, especially in the context of an aging population, which is characterized by a high cardiovascular mortality. A derangement of sulfur amino acid metabolism is de rigueur; in particular moderately high plasma homocysteine and cysteine are present. Homocysteine and cysteine are utilized as substrates for H$_2$S biosynthesis by the enzymes cystathionine beta synthase (CBS) and CSE, generating lanthionine and homolanthionine as side products. The latter amino acids have therefore been proposed as stable and reliable biomarkers of H$_2$S production [1]. However, in CKD patients on hemodialysis, they represent typical retention products (without ruling out completely the possibility of an increased intestinal synthesis). They in fact accumulate in the plasma of hemodialysis patients and are partially removed by a single hemodialysis session. Lanthionine in particular has been proposed as a novel uremic toxin, able to hamper H$_2$S production in vitro and whose increase may be the cause of hyperhomocysteinemia in CKD [2, 3].

The determination of these molecules requires the application of methods with a high degree of specificity and sensitivity, on one hand, while avoiding high costs and lengthy analysis, on the other. Liquid chromatography-tandem mass spectrometry (LC-MSMS) methods have been successfully used in protocols to identify and determinate a wide range of metabolites in different matrices, as well as to study their metabolisms [4–6]. Thus, automatic LC-MSMS has become the technique of choice for several metabolite analysis, thanks to the capacity to simultaneously separate and determinate multi component mixtures. Modern multiplex instruments can analyze thousands of samples per month so that, notwithstanding the generally high instrumental costs, the cost of the individual assay is affordable [7]. In addition, the improved specificity and resolution offered by triple quadrupole mass spectrometry allow their application in profiling (namely, identification and determination) of metabolites in human biological matrices. The selectivity is achieved by the multiple reaction monitoring (MRM) characteristics of triple quadrupole mass spectrometry instruments. In MRM analysis, two stages of mass selection, i.e., selection of a precursor ion and monitoring of specific product ion(s) derived from the precursor, provide highest analytical specificity applicable to complex mixture analysis. Such a two-step selection allows greater signal-to-noise ratio necessary for the quantification of even low-expressing molecules [8].

Here, a method utilizing LC-MS/MS in MRM ion mode has been developed and evaluated for the determination of key H$_2$S metabolites in plasma.

2 Materials

Prepare all solutions using ultrapure water (prepared by purifying deionized water, to attain a sensitivity of 18 MΩ-cm at 25 °C) and analytical grade reagents. Prepare and store all reagents at −20 °C (unless indicated otherwise). Homocysteine (Hcy), homoserine (Hse), lanthionine (Lan), homolanthionine (Hla), and cystathionine (Cysta) are purchased from Sigma-Aldrich. All the solutions and solvents are of the highest available purity and are suitable for LC-MS analysis.

2.1 Preparation of Standard Solutions

Standard solutions are prepared at 1 mg/ml concentration in 0.1 M HCl in water. A stock solution of 500 pg\μl of each metabolite is used for optimization of the MRM transitions.

Standard solutions are prepared by serial dilution at the following concentrations: 0.5, 1, 5, 25, 50, 100, and 150 pg/μl.

All standards are kept at −20 °C before LC-MS/MS analysis.

3 Methods

Carry out all procedures at room temperature unless otherwise specified. Stock all solutions at −20 °C (unless indicated otherwise).

3.1 Spiked Samples

1. Perform blood withdrawal from uremic patients and healthy volunteers, utilizing Vacutainer SST II Advance (BD Diagnostics, silica clotact/gel).

2. Perform blood spinning to isolate the serum supernatant fraction from whole blood, by centrifugation at 500 × g for 15 min.

3. Prepare spikes by enriching blood samples with 500 pg/μl of Hcy, Cysta, Lan, Hla, and Hse.

4. Store spiked samples at −20 °C until the analysis.

3.2 Sample Preparation

1. Perform experiments in duplicate and average results.

2. Prepare 200 μl of serum sample for protein precipitation.

3. Perform protein precipitation by adding 600 μl of ethanol and vortexing thoroughly.

4. Store the mixture at −20 °C, for about 30 min, to complete protein precipitation.

5. Centrifuge the samples at 13,000 × g for 10 min.

6. Transfer the supernatant directly into HPLC auto sampler.

7. Analyze 1 μl of the mixture using LC-MS/MS.

3.3 LC-MS/MS Instrumentation and Conditions

1. Carry out LC-MS/MS analyses by using a 6420 triple Q system with a HPLC 1100 series binary pump (Agilent, Waldbronn, Germany).

2. Set up a Phenomenex Kinetex 5u 100 A C18 analytical column.

3. Use a mobile phase generated by mixing eluent A and B, where eluent A is 2% ACN and 0.1% formic acid and eluent B is 95% ACN and 0.1% formic acid.

4. Set the flow rate at 0.200 ml/min.

5. Set starting condition at 5–95% A in 8 min and then to 100% for 2 min.

6. Perform tandem mass spectrometry using a turbo ion spray source operated in positive mode; the MRM mode is used for the selected analytes.

7. Use a standard solution of 500 pg\μl of each metabolite for optimization of the MRM transition.

8. Choose the ideal conditions for detection by means of the Agilent MassHunter Optimizer software.

9. Look at mass spectral parameters reported in Table 1.

3.4 Data Processing

Choose the ideal conditions for transitions detection via the Agilent MassHunter Optimizer software. Extracted mass chromatogram peaks of metabolites are integrated using Agilent MassHunter Quantitative Analysis software (B.05.00).

3.5 Method Validation

3.5.1 Limit of Detection and Limit of Quantitation

Consider the following definitions. The limits of detection (LODs) are defined as the lower limit of concentration below which the sample cannot be revealed and are determined by making ten replies of blank samples spiked with low concentrations of

Table 1
Precursor ions, product ions, and mass spectral parameters optimized for all analytes

Compound name	Precursor ion *m/z*	Product ion *m/z*	DP voltage (V)	Collision energy (V)
Cysta	223.3	134.1	80	5
	223.3	88		25
Lan	209.3	120	80	9
	209.3	74		25
Hcy	136.2	90	80	5
	136.2	56.1		17
Hse	120.1	74	80	5
	120.1	56.1		17
Hla	237.1	148	80	5
	237.1	102		30

Table 2
Validation parameters of the analytical method developed for the quantitative determination of Cys, Lan, Hcy, Hla, and Hse

Compound name	Linear range	R^2	Recovery %	LOD (pg/μl) (limit of detection)	LOQ (pg/μl) (limit of quantification)
Cystathionine	$y = 29.12x + 20.99$	0.998	90	0.54	1.62
Lanthionine	$y = 409.9x - 225.7$	0.997	94	0.38	1.15
Homocysteine	$y = 120.5x + 1050.1$	0.961	92	0.31	0.93
Homoserine	$y = 263.1x + 2650.1$	0.981	91	0.14	0.43
Homolanthionine	$y = 156.3x + 443.1$	0.994	94	0.42	1.07

each analyte. Calculations are made according to the following formula:

$$LOD = 3 \times STD \ (STD = \text{standard deviation}).$$

The limit of quantitation (LOQ) is the lowest concentration at which the analyte cannot only be reliably detected but also quantitated. The LOQ may be equivalent to the LOD or higher. Standard solutions are prepared by spiking a known amount of Hcy, Cysta, Lan, Hla, and Hse in the control serum samples. The metabolites are extracted and analyzed by the LC-MS/MS procedure (Table 2).

3.5.2 Matrix Effect

Evaluate possible matrix effects by comparing standard and matrix-matched calibration curves for each analyte. Prepare standard solutions as described. Repeat calibration curves three times.

3.5.3 Specificity

Check assay specificity by testing for interfering peaks at the retention time of the target analytes.

3.5.4 Stability

Determine stability of analytes in matrices on short-term conditions (−20 °C, 7 days). Evaluate all sample stability studies by comparing the area of the analytes in freshly prepared matrices with that of storage matrices and perform in triplicate. When the area of the analytes in storage matrices is lower than 10% compared to the area of the analytes in freshly prepared matrices, the analytes in matrices are considered unstable.

3.5.5 Selectivity

Evaluate selectivity by analyzing 20 blank samples. Prepare blank samples by executing the whole analysis procedure without test sample and omitting the addition of standards.

Consider selectivity acceptable when no peaks are detected in the chromatogram of the procedural blank sample at the retention time of the analytes ±0.1 min or, if present, peaks do not exceed 30% of the height of the native analyte in the chromatogram of the lowest calibration level.

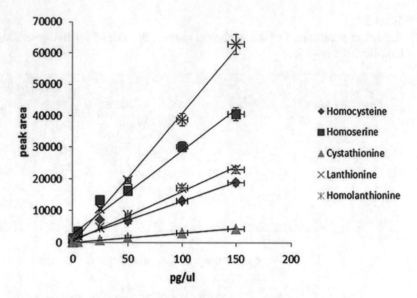

Fig. 1 Linear relationship between analyte concentration and peak area

3.5.6 Linearity

Take into account that the linearity of an analytical procedure is defined as its ability to obtain test results that are directly proportional to the concentration (amount) of the analyte in the sample within a given range. Linearity is demonstrated directly on the standard solutions by dilution of a standard stock solution as described. Linearity is determined by a series of three injections of standards at different concentration levels.

Obtain calibration curves using external standard method by plotting peak areas against the relevant concentration (pg/µl), and apply linear functions to the calibration curves. Integrate data by MassHunter quantitative software, showing a linear trend in the calibration range for all molecules (Fig. 1).

3.5.7 Precision (Repeatability and Intermediate Precision)

Assess precision at the lowest limit of the working range level from matrix samples, which are spiked with a known amount of standard solutions. Perform triplicate analyses on three different days on each spiked sample.

Estimate repeatability and intermediate precision by ANOVA calculation and express the result as relative standard deviation (RSDr). Evaluate intermediate precision from triplicate analysis in three independent analysis sequences performed over a period of 1 month.

3.5.8 Recovery

Calculate recovery values from the measurement of spiked standards in test samples. The average recoveries and their relative standard deviations are calculated from all the results obtained for each standard via response factors, determined from calibration solutions as follows:

$$\text{Recovery } (\%) = \frac{c_1 - c_2}{c_3} \times 100$$

where

c_1: analyte concentration measured after the addition.

c_2: analyte concentration measured before the addition.

c_3: added concentration.

Recovery values are not used for correction of analysis results. They are only used to monitor the yield of the sample preparation procedures (Table 2).

3.5.9 Measurement Uncertainty

Estimate measurement uncertainty according to the law of error propagation. The uncertainty contributions considered in the combined uncertainty are the uncertainties of the preparation of native standard solutions for instrument calibration, the uncertainty of the preparation of spiking solutions for the preparation of test samples, the uncertainty contribution deriving from instrument calibration, the uncertainty coming from the precision of the analyses, and the uncertainty of bias.

4 Notes

1. The novelty of our method based on targeted mass spectrometry in multiple reaction monitoring ion mode lies in the determination of two important compounds of the H_2S pathway, Lan and Hla, while simultaneously analyzing Hcy, Hse, and Cys.

2. Samples are prepared by simple protein precipitation, thus avoiding long-lasting procedures.

3. Optimal conditions for metabolite extraction. To optimize the sample extraction procedure prior to LC-MS/MS, different parameters are evaluated: protein precipitation solvents, influence of sonication, and extraction time. After adding to a control plasma the standards solution, standards are extracted in the supernatants by using three different solvents for protein precipitation. Protein precipitation with ethanol (1:3 v/v) at −20 °C for about 30 min resulted in the highest yields for protein pellet and for all analytes. Extending the precipitation time beyond 30 min (i.e., up to 2 h) does not increase analyte recovery. Moreover, this procedure is compatible with the mobile phase for further LC; thus, this solvent is subsequently used throughout the study.

4. No interference peaks and carryover peaks (i.e., signal below LOD) are observed in the blank samples after the high

calibrator is injected. The ion suppression effects are evaluated, resulting in no significant ion suppression (i.e., lower than 10%) of the analytes of interest.

5. To assess the applicability of this method to routine clinical practice, we evaluated the stability of the metabolites of interest in serum over a 2-week period, which represents sufficient time for transporting the patient sample to the laboratory.

6. The stability is tested by analyzing serum samples spiked with target molecules at final concentrations of 500 pg/μl. All the samples are stored at −20 °C and the MRM analyses showed a decrease in concentrations of 10%, after 7 days of storage. Because these decreases in the analyte concentrations are lower than the inter-assay variability, they do not appear to be significant.

7. The analytical performance of our method, i.e., its linearity, recovery, and reproducibility, is analogous to the performance previously described in other studies based on the detection of metabolites in plasma matrix, indicating the suitability of our method for routine use.

8. In the newly developed method, the LOQ are sufficiently lower than their physiological concentrations, whereas the LOQ of Cysta is at the lower limit of the reference range. The limited sensitivity of our method may therefore cause problems in determining Cysta at the lower limit of the reference range in the blood. Moreover, severe alterations of the hematocrit may also influence the concentration of analytes and should be considered. Despite these limitations, the method is suitable for identifying patients with moderate to severe hyperhomocysteinemia and renal failure.

9. Low reagent costs together with the relative simplicity of sample preparation make the LC-MS/MS method well suited, not only for research work but also in those laboratories with a tandem mass spectrometer, for the measurement of routine clinical samples.

10. Once set up, this MRM method can be successfully applied as a rapid screening procedure for detecting the abovementioned analytes in patients with hyperhomocysteinemia and renal failure.

References

1. Chiku T, Padovani D, Zhu W, Singh S, Vitvitsky V, Banerjee R (2009) H$_2$S biogenesis by human cystathionine gamma-lyase leads to the novel sulfur metabolites lanthionine and homolanthionine and is responsive to the grade of hyperhomocysteinemia. J Biol Chem 284 (17):11601–11612

2. Perna AF, Zacchia M, Trepiccione F, Ingrosso D (2017) The sulfur metabolite lanthionine: evidence for a role as a novel uremic toxin. Toxins

(Basel) 9(1):pii: E26. https://doi.org/10.3390/toxins9010026

3. Perna AF, Di Nunzio A, Amoresano A, Pane F, Fontanarosa C, Pucci P, Vigorito C, Cirillo G, Zacchia M, Trepiccione F, Ingrosso D (2016) Divergent behavior of hydrogen sulfide pools and of the sulfur metabolite lanthionine, a novel uremic toxin, in dialysis patients. Biochimie 126:97–107

4. Dettmer K, Aronov PA, Hammock BD (2007) Mass spectrometry-based metabolomics. Mass Spectrom Rev 26(1):51–78

5. Kitteringhama NR, Jenkinsa RE, Laneb CS, Elliott VL, Park BK (2009) Multiple reaction monitoring for quantitative biomarker analysis. J Chromatogr B 877:1229–1239

6. Bártl J, Chrastina P, Krijt J, Hodík J, Pešková K, Kožich V (2014) Simultaneous determination of cystathionine, total homocysteine, and methionine in dried blood spots by liquid chromatography/tandem mass spectrometry and its utility for the management of patients with homocystinuria. Clin Chim Acta 437:211–217

7. Gosetti F, Mazzucco E, Gennaro MC, Marengo E (2013) Ultra high performance liquid chromatography tandem mass spectrometry determination and profiling of prohibited steroids in human biological matrices. A review. J Chromatogr B 927:22–36

8. Prasad B, Unadkat DJ (2014) Optimized approaches for quantification of drug transporters in tissues and cells by MRM proteomics. AAPS J 16(4):634–648. Published online 2014 April 22. https://doi.org/10.1208/s12248-014-9602

Chapter 3

Analysis of Vascular Hydrogen Sulfide Biosynthesis

Thomas J. Lechuga and Dong-bao Chen

Abstract

With potent vasodilatory and pro-angiogenic properties, hydrogen sulfide (H_2S) is now accepted as the third gasotransmitter after nitric oxide (NO) and carbon monoxide. Endogenous H_2S is mainly synthesized by cystathionine β-synthase (CBS) and cystathionine γ-lyase (CSE). Akin to previous studies showing hormonal regulation of NO biosynthesis, we first reported that uterine and systemic artery H_2S biosynthesis is regulated by exogenous estrogens in an ovariectomized sheep model of estrogen replacement therapy, specifically stimulating CBS, but not CSE, expression, in uterine (UA) and mesenteric (MA), but not carotid (CA), arteries in ovariectomized nonpregnant sheep. We have found significantly elevated H_2S biosynthesis due to CBS upregulation under estrogen-dominant physiological states, the proliferative phase of menstrual cycle and pregnancy in primary human UAs. Our studies have pioneered the role of H_2S biology in uterine hemodynamics regulation although there is still much that needs to be learned before a thorough elucidation of a role that H_2S plays in normal physiology of uterine hemodynamics and its dysregulation under pregnancy complications can be determined. In this chapter we describe a series of methods that we have optimized for analyzing vascular H_2S biosynthesis, including (1) real-time quantitative PCR (qPCR) for assessing tissue and cellular levels of CBS and CSE mRNAs, (2) immunoblotting for assessing CBS and CSE proteins, (3) semiquantitative immunofluorescence microscopy to specifically localize CBS and CSE proteins on vascular wall and to quantify their cellular expression levels, and (4) methylene blue assay for assessing H_2S production in the presence of selective CBS and CSE inhibitors.

Key words H_2S, qPCR, Immunoblotting, Immunofluorescence microscopy, Methylene blue assay

1 Introduction

Hydrogen sulfide (H_2S) has now been accepted as the third "gasotransmitter" after nitric oxide (NO) and carbon monoxide due to its NO-like biological properties [1–3]. Endogenous H_2S is mainly synthesized by two key enzymes, cystathionine β-synthase (CBS) and cystathionine γ lyase (CSE) [4, 5]. These enzymes produce H_2S from L-cysteine, CBS via a β-replacement reaction with a variety of thiols and CSE by disulfide elimination followed by reaction with various thiols [4, 5]. H_2S can be also synthesized via 3-mercaptopyruvate sulfurtransferase (3MST), cytosolic cysteine aminotransferase (cCAT), and mitochondrial cysteine

Jerzy Bełtowski (ed.), *Vascular Effects of Hydrogen Sulfide: Methods and Protocols*, Methods in Molecular Biology, vol. 2007,
https://doi.org/10.1007/978-1-4939-9528-8_3, © Springer Science+Business Media, LLC, part of Springer Nature 2019

aminotransferase (mCAT), but to a much less extent [6]. In mammals, H_2S potently dilates various vascular beds via activating ATP-dependent potassium (K_{ATP}) channel [7] and relaxes smooth muscle via activating large conductance calcium-activated potassium (BK_{Ca}) channel [8]. H_2S also promotes angiogenesis in vitro and in vivo [9]. Thus, H_2S functions as a potent vasodilator [10].

Estrogens are potent vasodilators that cause blood flow to rise in organs throughout the body with the greatest effects occurring in reproductive tissues, especially the uterus [11, 12]. In ovariectomized (OVX) nonpregnant ewes, daily estradiol-17β (E_2) increases basal uterine blood flow by 30–40% after 6–7 days in the absence of changes in arterial pressure or heart rate [13] and reduced responses to vasoconstrictors [14]. In addition, acute E_2 exposure causes even more robust up to tenfold rise in UBF within 90–120 min after a bolus intravenous injection of 1 μg/kg [15, 16]. This uterine vasodilatory effect of estrogens is of major physiological significance during the follicular phase of the ovarian cycle and pregnancy in which circulating estrogens are significantly elevated [16–19]. During normal pregnancy, estrogen levels increase substantially to upregulate UBF that provides the sole source of nutrients and oxygen supplies for the fetus and the exit of the metabolic wastes and respiratory gases of the fetus. UBF is a critical rate-limiting factor for a healthy pregnancy because (1) dramatic rise in uterine blood flow in the last one-third of gestation is directly linked to fetal growth and survival and (2) insufficient rise in uterine blood flow results in preeclampsia, intrauterine growth restriction, and many other pregnancy disorders, affecting late life after birth and the mother's well-being during pregnancy and postpartum [20]. Thus, mechanistic investigations on estrogen-induced and pregnancy-associated uterine and systemic vasodilatation will identify therapeutic targets or even preventive options for pregnancy disorders.

We have reported that exogenous estrogen replacement therapy stimulates CBS, but not CSE, mRNA and protein expression, and promotes H_2S production in uterine (UA) and mesenteric (MA) but not carotid (CA) arteries [21]. Additionally, we have found that CBS, but not CSE, is upregulated in both the endothelium and smooth muscle of human UA during the estrogen "dominant" physiological states, i.e., the proliferative phase of the menstrual cycle and pregnancy [22]. Other enzymes, i.e., 3MST, cCAT, and mCAT, are also detectable in human UA but unchanged during the menstrual cycle and pregnancy. In organ bath studies using wire myography, we also have found that a slow-releasing H_2S donor GYY4137 potently dilates primary UAs of nonpregnant and pregnant rats; however, the sensitivity (pD2) to GYY4137 is significantly greater ($p < 0.001$) in pregnant (7.43 ± 0.02) than nonpregnant (5.97 ± 0.01) UA. Moreover, GYY4137 does not dilate pregnant rat MAs [23]. Thus, these data suggest that augmented H$_2$S via selective upregulation of endothelial and smooth muscle CBS plays a role in estrogen-stimulated and pregnancy-augmented vasodilation in UA and selected vascular beds.

In this chapter, we described a series of methods that we have developed for analyzing vascular tissue and cell H$_2$S biosynthesis, including (1) real-time quantitative PCR (qPCR) for assessing steady-state levels of CBS and CSE mRNAs, (2) immunoblotting for assessing CBS and CSE proteins, (3) semiquantitative immunofluorescence microscopy for quantifying CBS and CSE proteins and their cellular (endothelium *vs.* smooth muscle) localization, and (4) methylene blue assay for assessing H$_2$S production in the presence of selective CBS and CSE inhibitors.

2 Materials

Prepare all solutions without sodium azide using ultrapure water (prepared by purifying deionized water, to attain a sensitivity of 18 MΩ-cm at 25 °C) and analytical grade reagents. Prepare and store all reagents at room temperature unless indicated. Diligently follow all waste disposal regulations when disposing waste materials.

2.1 CBS/CSE mRNA Analysis

2.1.1 RNA Purification and cDNA Synthesis

1. TRIzol reagent (Life Technologies, Carlsbad, CA, USA) or similar acid-guanidinium-phenol reagent.
2. Fine microdissection scissors.
3. DNase/RNase-free 1.7 mL microcentrifuge tubes.
4. Benchtop microcentrifuge.
5. Direct-zol RNA Mini Prep (Zymo Research, Irvine, CA, USA).
6. Biowave DNA (Biochrom, Cambridge, UK).
7. Random primers (Promega, Madison, WI, USA).
8. Nuclease free water.
9. Thermocycler.
10. 10 mM dNTP (Promega).
11. Moloney Murine Leukemia Virus Reverse Transcriptase (M-MLV RT; Promega).
12. M-MLV 5× reaction buffer (Promega)

2.1.2 qPCR

1. Primers (*see* Table 1).
2. RT^2 SYBR Green/ROX qPCR master mix (Qiagen, Germantown, MD, USA).
3. 96-well optical PCR plate.
4. PCR optical adhesive.
5. StepOnePlus real-time PCR system (Applied Biosystems, Foster City, CA, USA).

Table 1
Primers used for qPCR analysis of CBS and CSE mRNAs

Gene	Forward	Reverse	Amplicon size
CBS	5'-TGAGATTGTGAGGACGCGCCCAC-3'	5'-TCACACTGCTGCAGGATCTC-3'	177 bp
CSE	5'-TTGTATGGATGATGTGTATGGAAGG-3'	5'-CCAAACAAGCTTGGTTTCTGGTG-3'	141 bp
L-19	5'-AGACCCCAATGAGAGACCAATG-3'	5'-GTGTTTTTCCGGCATCGAGC-3'	129 bp

2.2 CBS/CSE Protein Analysis

2.2.1 Sample Lysate Preparation

1. Fine microdissection scissors.

2. Nondenaturing lysis buffer containing protease inhibitor: 10 mM Tris–HCl, pH 7.4, 100 mM NaCl, 1 mM EDTA, 1 mM EGTA, 1% Triton X-100, 0.5% NP-40, 50 mM NaF, 1 mM Na$_3$VO$_4$, 1 mM PMSF, 1% protease inhibitor cocktail.

3. Sonicator.

4. Benchtop microcentrifuge.

5. Pierce BCA Protein Assay Kit (Life Technologies, Grand Island, NY, USA).

6. 5× SDS-PAGE sample buffer: 250 mM Tris–HCl pH 6.8, 10% (w/v) SDS, 30% (v/v) glycerol, 0.02% (w/v) bromophenol blue, 5% (v/v) β-mercaptoethanol.

2.2.2 SDS-PAGE and Transfer

1. Acrylamide gel casting apparatus and combs.

2. Distilled water.

3. 4× Resolving Gel buffer: 1.5 M Tris, 0.4% SDS, pH 8.8.

4. Acrylamide/Bis (30%/29:1).

5. 10% (w/v) ammonium persulfate (APS).

6. 50% glycerol.

7. N,N,N',N'-tetraethylmethylenediamine (TEMED).

8. Resolving gel: 6.4 mL H$_2$O, 5 mL 4× resolving gel buffer, 6.6 mL 30% (v/v) acrylamide/bis, 5 mL, 2 mL 50% glycerol, 10 µL of TEMED, and 100 µL of 10% (w/v) APS (*see* **Note 1**).

9. 4× Stacking Gel buffer: 0.5 M Tris, 0.4% SDS, pH 6.8.

10. Stacking gel: 12 mL H$_2$O, 5 mL 4× resolving gel buffer, 3 mL 30% (v/v) acrylamide/bis, 10 µL of TEMED, and 400 µL of 10% (w/v) APS.

11. 20 × 10 cm gel system containing loading rig, glass plates, and buffer tank.

12. SDS-PAGE pre-stained polypeptide markers (Thermo Scientific, Waltham, MA, USA).

13. SDS-PAGE running buffer: 25 mM Tris, 192 mM glycine, 0.1% (w/v) SDS.

14. Nitrocellulose membrane.

15. Methanol.

16. Blot Absorbent Filter Paper.

17. Transfer buffer.

18. Semidry transfer system.

2.2.3 Immunoblotting

1. Benchtop rocker.

2. Blocking solution.

3. Mouse anti-CSE primary antibody (S-374249, Santa Cruz Biotech, Dallas, TX, USA).

4. Rabbit anti-CBS primary antibody (ab131155, Abcam, Cambridge, UK).

5. TBS-T: 20 mM Tris, 150 mM NaCl, 0.1% Tween-20, pH 7.5.

6. Anti-rabbit HRP-conjugated secondary antibody (Cell Signaling, Beverly, MA, USA).

7. Anti-mouse HRP-conjugated secondary antibody (Cell Signaling).

8. TBS: 20 mM Tris, 150 mM NaCl, pH 7.5.

9. Clear film (such as standard kitchen plastic wrap).

10. SuperSignal® West Femto Maximum Sensitivity Substrate (Thermo Fisher Scientific, Waltham, MA).

11. ChemiImager Imaging System (Alpha Innotech, San Leandro, CA, USA).

12. Mild stripping buffer: 15 g glycine, 0.1% SDS, 1% Tween-20, 1 L H_2O, pH 2.2.

13. PBS (1×): 137 mM NaCl, 2.7 mM KCl, 10 mM Na_2HPO_4, 1.8 mM KH_2PO_4.

14. Rabbit anti-β-actin (Ambion, Austin, TX, USA).

15. NIH Image J (https://imagej.nih.gov/ij/).

16. Microsoft Excel (Microsoft Corp, Redmond, WA, USA).

2.3 Semiquantitative Fluorescence Microscopy

1. Slides containing tissue sections (*see* **Note 2**).

2. Xylenes.

3. Ethanol (100%, 90%, 70%).

4. Deionized water (diH_2O).

5. 1× PBS: 137 mM NaCl, 2.7 mM KCl, 10 mM Na_2HPO_4, 1.8 mM KH_2PO_4.

6. 0.05% Trypsin (Thermo Fisher Scientific).

7. 300 mM Glycine-PBS: 1 L diH_2O, 22.56 g glycine, filtered (*see* **Note 3**).

8. Blocking solution: 50 mL 1× PBS, 50 μL gelatin, 0.5 g BSA, 0.0625 g saponin.

9. PAP pen (Invitrogen, Carlsbad, CA, USA).

10. Mouse anti-CD31 primary antibody (Dako, Carpinteria, CA, USA).

11. Antibody dilution buffer: 50% blocking solution, 50% 1× PBS.

12. Alexa 568 conjugated goat anti-mouse IgG secondary antibody.

13. Mouse anti-CSE primary antibody (Santa Cruz Biotech).

14. Rabbit anti-CBS primary antibody (Abcam).

15. Alexa 488 conjugated goat anti-mouse IgG secondary antibody (Invitrogen).

16. Alexa 488 conjugated donkey anti-rabbit IgG secondary antibody (Invitrogen).

17. ProLong Gold Antifade Mountant with DAPI (Invitrogen).

18. Coverslips.

19. Leica fluorescence microscope (Leica Corp, Deerfield, IL, USA).

20. Charge-coupled device camera with the SimplePCI image analysis software (Hamamatsu Corp, Sewickley, PA, USA).

21. Microsoft Excel (Microsoft Corp).

2.4 H$_2$S Activity Assay

1. Fine microdissecting scissors.

2. Sonicator.

3. Benchtop microcentrifuge.

4. Pierce BCA Protein Assay Kit (Life Technologies, Grand Island, NY, USA).

5. 37 °C water shaking water bath.

6. 12 mL outer tubes.

7. 2 mL inner tubes.

8. 0.5 × 1.5 cm cut filter paper.

9. 2.5 × 2.5 cm cut parafilm squares.

10. 50 mM potassium phosphate buffer, pH 8.0.

11. 200 mM L-cysteine.

12. 80 mM pyridoxal 5′-phosphate.

13. β-cyano-L-alanine (BCA).

14. *O*-(carboxymethyl)hydroxylamine hemihydrochloride (CHH).

15. 1% zinc acetate.

16. 50% trichloroacetic acid (TCA).

17. *N*,*N*-dimethyl-*p*-phenylenediamine sulfate, 20 mM in 7.2 M HCl.

18. FeCl$_3$, 30 mM in 1.2 M HCl.

19. Sodium hydrosulfide (NaHS).

20. Deionized water.

3   Methods

3.1   Analysis of CBS and CSE mRNAs

3.1.1   RNA Extraction

1. Finely mince ~1 cm length of snap-frozen artery tissues (~100 mg) in 600 μL ice-cold TRIzol reagent (Life Technologies, Carlsbad, CA, USA, #15596) with fine microdissecting scissors (Fisher Scientific, Hampton, NH) in a 1.7 mL microcentrifuge tube (Fisher) on ice (*see* **Note 1**).

2. Remove tissue debris by benchtop centrifugation at $12,000 \times g$ for 1 min; transfer supernatant to a fresh RNase-free tube (Fisher).

3. Proceed to RNA purification using Direct-zol RNA Mini Prep (Zymo Research, Irvine, CA, USA, R2052), following manufacturer's protocol and elute purified RNA in 30 μL of nuclease-free water.

4. Dilute 2 μL (1:100) of RNA in 198 of nuclease-free water to quantify RNA by $OD_{260/280}$.

3.1.2   cDNA Synthesis: Reverse Transcription

1. Add 1 μg purified RNA, 1 μL random primers (Promega, Madison, WI, USA), and nuclease-free water up to 14 μL in RNase-free microcentrifuge tube.

2. Heat the tube for 5 min at 70 °C on a thermocycler to melt template secondary structure.

3. Immediately return on ice to prevent template secondary structure from reforming.

4. Add 4 μL of 5× RT reaction buffer (Promega), 1 μL of 10 mM dNTPs (Promega), and 200 units (1 μL) of Moloney Murine Leukemia Virus Reverse Transcriptase (Promega) for 20 μL total reaction volume. Gently mix.

5. Incubate the tube for 1 h at 37 °C.

6. Return to ice and dilute cDNA template with 80 μL of nuclease-free water (final volume = 100 μL).

3.1.3   qPCR

1. Prepare 10 μM stocks for each set of primers of CBS, CSE, and L-19 (Table 1).

2. Prepare qPCR master mix. For each reaction, mix 7.5 μL of 2× RT^2 SYBR Green/ROX qPCR master mix (Life Technologies, #4367660), 3.6 μL nuclease-free H_2O, and 0.45 μL of 10 μM primer sets for each target.

3. Add 3 μL of cDNA template into a well of an optical 96-well qPCR plate (Life Technologies, 4346906). Run each sample in triplicate. Seal the plate with PCR optical adhesive film.

4. Quick-spin plate to sediment the cDNA-master mix reaction to the bottom of the well.

5. StepOnePlus real-time PCR system (Applied Biosystems, #4376592). Set up computer to run plate with SYBR Green for quantitative comparative ΔΔCt.

6. Assign targets for appropriate primer pair (CBS, CSE, or L19) and sample designation.

7. Set up the following real-time thermal cycler program: 95 °C for 10 min, followed by 40 cycles of 95 °C for 15 s, 55 °C for 30 s, and 72 °C for 30 s.

8. Program the StepOnePlus real-time PCR system to turn on optics to detect SYBR green signal after the 72 °C extension step of each cycle.

9. Set the StepOnePlus real-time PCR system to run the following melting curve program immediately after completion of the above program: 95 °C for 15 s, 60 °C for 1 min (optics off), and 60 °C to 95 °C at 2 °C per min (optics on).

10. Calculate by subtracting the average L19 Ct for a triplicate from each Ct for CBS or CSE of that sample and average in Microsoft excel.

11. Calculate the $2^{-\Delta\Delta Ct}$ relative mRNA levels of each triplicate measurement for each sample by taking the log base −2 of each Ct (*see* **Note 2**).

12. Average the triplicate and calculate fold changes vs. the average of the control group (Fig. 1).

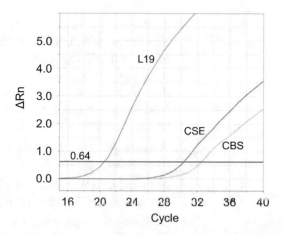

Fig. 1 qPCR amplification plots for CBS, CSE, and L19 in arteries. Basal levels of CBS, CSE, and L19 mRNAs were determined by real-time qPCR in uterine (UA), mesenteric (MA), and carotid (CA) arteries. A typical mRNA amplification plot for each mRNA is shown. Ct value was determined by the default Ct threshold (0.64) using a StepOnePlus real-time PCR system (Applied Biosystems) and accompanied software. The average Ct value for CBS was approximately 33, 30 for CSE, and 33 for L19

3.2 Immunoblot Analysis of CBS and CSE Proteins

3.2.1 Protein Extraction and Sample Preparation

1. Finely mince ~1.5 cm length of artery tissues in 500 μL ice-cold nondenaturing lysis buffer [24] with protease inhibitor cocktail (Fisher) with fine microdissecting scissors in a 1.7 mL micro-centrifuge tube on ice (*see* **Note 3**).

2. Sonicate samples for 30 s.

3. Spin samples in a benchtop microcentrifuge at 12,000 RPM (~13,500 × *g*) at 4 °C for 10 min.

4. Transfer supernatant to a fresh microcentrifuge tube and repeat **step 3**.

5. Determine protein concentration of lysates using the Pierce BCA Protein Assay Kit (Life Technologies) following the manufacturers protocol.

6. Mix the protein samples with 5× SDS-PAGE sample buffer [250 mM Tris–HCl pH 6.8, 10% (w/v) SDS, 30% (v/v) glycerol, 0.02% (w/v) bromophenol blue, 5% (v/v) β-mercaptoethanol] and use 1× sample buffer to bring up the protein concentration to 1 μg/μL. Boil the samples for 5 min at 95 °C on a heat block.

7. Quick spin samples and keep on ice until use or store at −20 °C for later use.

3.2.2 SDS-PAGE and Transfer

1. Load 20 μg (e.g., 20 μL of 1 μg/μL) of protein samples into the wells of a 10% SDS-PAGE mini-gel; load 5 μL of pre-stained polypeptide marker (Thermo Scientific, Waltham, MA, USA, 26616) into the first well and 2.5 μL in the last well (*see* **Note 4**).

2. Secure the gel in an vertical mini-gel system (C.B.S. Scientific, San Diego, CA, USA, #MGV-102) and initiate electrophoresis at 80 V until the blue dye in the samples have passed the interface of the stacking/resolving gels, and then run at 120 V until the blue dye in the samples has nearly ran to the bottom of the gel. Check occasionally and add running buffer to top reservoir if needed.

3. Remove gel from glass plates and trim the stacking gel and any unnecessary gel (*see* **Note 5**).

4. Soak gel in transfer buffer (25 mM Tris 190 mM glycine 20% methanol, pH 7.5) for 15 min on a benchtop rocker.

5. Activate a nitrocellulose membrane (Thermo Scientific, #88518) with methanol, and incubate in transfer buffer for 15 min with gentle rocking.

6. In a large trey with transfer buffer, sandwich the inverted gel with a nitrocellulose membrane on top with two filter papers. Complete the sandwich with two additional filter papers. Avoid any air bubbles in the sandwich.

7. Moisten the surface of the top and bottom plates of a semidry transfer system (Thermo Scientific, #HEP-3) with transfer buffer.

8. Place sandwich on semidry transfer system and carefully cover with the top plate. Lock and tighten the system evenly.

9. Run transfer at 200 mA per sandwich for 1.5 h.

10. Open the semidry transfer system. Remove top filter paper. Remove and invert the membrane and then orient it by cutting the top right corner. Soak the membrane in TBS buffer (20 mM Tris, 150 mM NaCl, pH 7.5) (*see* **Note 6**).

3.2.3 Immunoblotting

1. Gently remove gel, invert, and stain with Coomassie blue if desired (*see* **Note 7**).

2. Transfer membrane to 5% BSA in TBS-T (TBS + 0.1% Tween-20) blocking solution for 1 h.

3. Seal the membrane in a plastic pouch with primary antibodies diluted in 2 mL of 5% BSA in TBS-T. Store at 4 °C overnight with agitation.

4. Rabbit ant-CBS polyclonal antibody (Abcam ab131155, 1:200 for 0.2 μg/mL final concentration). CBS will be detected at approximately 61 kDa. Mouse anti-CSE monoclonal antibody (Santa Cruz S-374249, 1:500 for 0.4 μg/mL final concentration). CSE will be detected at approximately 45 kDa.

5. Wash the membrane with TBS-T for 3 × 5 min on a rocker.

6. Incubate the blot with HRP-linked anti-rabbit IgG (Cell Signaling, Beverly, MA, USA #7074) for CBS or HRP-linked anti-mouse IgG (Cell Signaling #7076) at 1:1000 in TBS-T for 1 h at room temperature.

7. Wash the membrane with TBS-T for 3 × 5 min on a rocker.

8. Incubate the membrane with the SuperSignal® West Femto Maximum Sensitivity Substrate (Fisher #34095); typically less than 1 mL is needed per blot.

9. Visualize and capture chemiluminescence signals in Alpha Innotech ChemiImager Imaging System. Save the images at 8-bit Tagged Image File Format (TIFF) files for analysis.

10. Strip the membrane by incubation in a mild stripping buffer (0.1% SDS, 0.15% glycine, 1% Tween20, pH 2.2) at 50 °C for 45 min with some agitation.

11. Re-probe the membrane with rabbit anti-β-actin (Ambion AM4302, 1:10,000 for 0.1 μg/mL final concentration) at 4 °C overnight with agitation. β-Actin will be detected at approximately 42 kDa.

12. Wash the membrane with TBS-T for 3 × 5 min on a rocker.

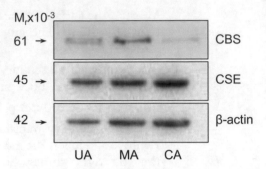

$M_r x10^{-3}$

61 →   CBS

45 →   CSE

42 →   β-actin

UA   MA   CA

Fig. 2 Immunoblots of CBS, CSE, and β-actin in arteries. CBS, CSE, and β-actin proteins were determined by immunoblotting in uterine (UA), mesenteric (MA), and carotid (CA) arteries. 20 μg of lysates were ran on a 10% acrylamide gel. The antibodies detected a band of approximately 61 kDa for CBS, 45 kDa for CSE, and 42 kDa for β-actin

13. Incubate the blot with HRP-linked anti-rabbit IgG (Cell Signaling #7074, 1:1000).

14. Repeat **steps 7–9** to detect β-actin signals and save the images.

15. Use NIH Image J to conduct densitometry analysis of CBS, CSE, and β-actin.

16. Normalize each CBS/CSE band to corresponding β-actin signals (Fig. 2).

3.3 Semiquantitative Immunofluorescence Microscopy

3.3.1 Immunofluorescence Microscopy

1. Fix 2 cm freshly collected artery rings in 4% paraformaldehyde overnight and embedded in paraffin, and then cut 5 μm sections and place on glass slides.

2. Place slides in a slide rack and de-paraffinize sections in 100% xylene 2 × 10 min.

3. Hydrate sections in graded ethanol (100%, 95%, 70%), each grade for 2 × 5 min.

4. Wash slides 1 × 5 min in diH_2O and then 1 × 5 min in 1× PBS.

5. Antigen retrieval: incubate slides with pre-warmed (37 °C) 0.05% Trypsin (Fisher # 25300054) at RT for 30 min, and then wash slide in deionized water (diH_2O) for 5 min.

6. Incubate the slides in 300 mM glycine in PBS (137 mM NaCl, 2.7 mM KCl, 10 mM Na_2HPO_4, 1.8 mM KH_2PO_4, 0.45 μM filtered; *see* **Note 8**) for 3 × 20 min at RT to quench autofluorescence.

7. Block non-specific proteins in PBS blocking solution containing 0.01% gelatin, 1% BSA, and 0.001% saponin at RT for 1 h.

8. Incubate the slides with anti-CD31 antibody (Dako #M0823, 1:40) in antibody dilution buffer (50% blocking solution and 50% 1× PBS) in humidifying chamber at 4 °C overnight (*see* **Note 9**).

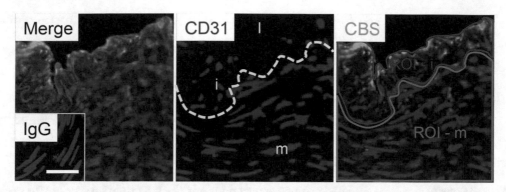

Fig. 3 Semiquantitative assessment and localization of CBS and CSE. UA sections were labeled with antibodies against CBS, or CSE (not shown), followed by secondary Alexa[488] (green)-labeled secondary antibody. Endothelial cells were labeled with CD31 followed by Alexa[568] (red)-labeled secondary antibody. Cell nuclei were stained with DAPI. Negative control treated with IgG or with primary antibody omitted is shown in first panel. Representative outline of border between UA intima and media was indicated, and UA lumen (l), intima (i), and media (m) were denoted (second panel). Regions of interest for intima (ROI – i) and media (ROI – m) used for analysis in Simple PCI software are illustrated (third panel). Scale bar is 25 μm

9. Wash slides 3 × 5 min in 1× PBS.

10. Incubate the slides with Alexa[568]-labeled goat anti-mouse IgG (Fisher A-11004, 1:1000) in humidifying chamber at RT for 1 h in dark (*see* **Note 10**).

11. Wash the slides 3 × 20 min in 1× PBS. Keep slides in dark (*see* **Note 11**).

12. Repeat **step 7**.

13. Incubate the slides with anti-CBS or CSE antibody (1 μg/mL) in antibody dilution buffer in humidifying chamber at 4 °C overnight.

14. Wash slides 3 × 5 min in 1× PBS.

15. Incubate the slides with Alexa[488] donkey anti-rabbit IgG (Thermo Scientific # A-21206 1:1000 for CBS) or Alexa[488]-labeled goat anti-mouse IgG (Thermo Scientific # A-11001, 1:1000 for CSE) in humidifying chamber at RT for 1 h in dark.

16. Repeat **step 11**.

17. Mount the slides with ProLong Gold Antifade Mountant with DAPI (Fisher P36930), and keep the slides overnight at RT in dark.

18. Visualize the slides with a fluorescence microscopy (20×) and capture the images in TIFF files (Fig. 3).

3.3.2   Image Analysis

1. Open image in Simple PCI Image analysis software (Hamamatsu Corp, Sewickley, PA, USA).

2. Calibrate image; we use 20× with 1.5× magnification settings.

3. Right click image and select "measurement ROI."

4. Select ROI tool and draw ROI on specific cell structures or cell types being assessed.

5. Select "measurements"; choose "area" and "mean green."

6. Choose "measure to spreadsheet."

7. Copy and paste the data to Excel.

8. Subtract average background fluorescence (from negative control images) from all images analyzed.

9. Report florescence as $RFI/\mu m^2$ (fold of control).

3.4 H_2S Activity Assay

3.4.1 Methylene Blue Assay

1. Finely mince artery segment (1–1.5 cm in length, ~100 mg) in ice-cold 50 mM potassium phosphate buffer, pH 8.0, with fine microdissecting scissors in 1.7 mL microcentrifuge tube on ice (*see* **Note 3**).

2. Sonicate samples for 30 s.

3. Spin samples in a benchtop microcentrifuge at 12,000 RPM for 10 min at 4 °C.

4. Transfer supernatant to a fresh microcentrifuge tube and repeat **step 3**.

5. Determine protein concentration of lysates using the Pierce BCA Protein Assay Kit.

6. Add the following reagents to a 12 mL outer tube:

   (a) Protein lysate (at least 40 μg protein) in 0.925 mL ice-cold potassium phosphate buffer.

   (b) 50 μL of L-cysteine (200 mM) to get 10 mM final concentration.

   (c) 25 μL of pyridoxal 5′-phosphate (80 mM) to get 2 mM final concentration.

7. Add O-(carboxymethyl)hydroxylamine hemihydrochloride (CHH, Sigma Aldrich #C13408) and/or β-cyano-L-alanine (BCA, Caymen Chemical, Ann Arbor, MI, USA, #10010947) for a final concentration of 2 mM to designated reaction tubes.

8. Soak a piece of filter paper (0.5 × 1.5 cm) with 0.3 mL of 1% zinc acetate in a 2 mL inner tube, and put the uncapped tube in each 12 mL outer tube.

9. Flush the tube with a slow stream of nitrogen gas (N_2) for 20 s and then seal with double layer of parafilm.

10. Transfer tubes to a 37 °C shaking water bath and incubate for 90 min.

11. Inject 0.5 mL of 50% trichloroacetic acid into the reaction mixture through the parafilm.

12. Incubate the tube in the shaking water bath for 60 min to complete the trapping of H_2S by the 1% zinc acetate solution to form zinc sulfide.

13. Remove the parafilm and add 50 μL of N,N-dimethyl-p-phenylenediamine sulfate (20 mM in 7.2 M HCl) and 50 μL of FeCl$_3$ (30 mM in 1.2 M HCl) to the inner 2 mL tube.

14. Measure OD$_{630}$ after 20 min incubation at RT.

3.4.2 Calibration Curve for Each Measurement

1. Freshly prepare sodium hydrosulfide (NaHS) stock solution (1 M) by dissolving 280 mg of NaHS into 5 mL of diH$_2$O.

2. Serial dilute NaHS stock solution into the following concentrations: 0, 7.8, 31.25, 62.5, 125, 250, 500, and 1000 μM (*see* **Note 12**).

3. Measure H$_2$S level as stated in 6.1 simultaneously.

4. Generate a H$_2$S calibration curve based on OD$_{630}$.

5. Use the calibration curve (Fig. 4) to determine H$_2$S in all unknown samples.

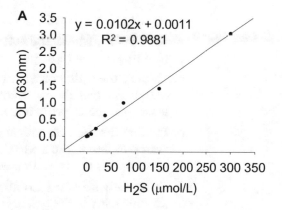

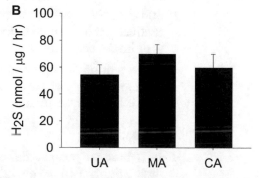

Fig. 4 Methylene blue assay to assess H$_2$S production in arteries. (**a**) Example of assay standard curve generated by serial dilutions of NaHS is shown. The linear equation formed by the curve should be used to determine the H$_2$S levels with the OD values produced from the unknown samples. (**b**) Basal levels of H$_2$S were determined by using the methylene blue assay in uterine (UA), mesenteric (MA), and carotid (CA) arteries

4 Notes

1. Due to limited quantities of arteries, we have chosen to finely mince the tissue. For RNA extraction it is more desirable to freeze the tissue in liquid N_2 and grind with a mortar and pestle. This technique can result in greater tissue loss; use only if excess tissue is available. Alternatively, total RNA can be harvested from in vitro studies using cell culture models (minimum 0.5×10^6 smooth muscle or endothelial cells).

2. Use the equation "=POWER(2,-X)" in Microsoft Excel, where X is each value generated in **step 10**.

3. Due to limited quantities of arteries, we have chosen to finely mince the tissue. For protein extraction it is more desirable to use a tissue homogenizer. This technique can result in greater tissue loss; use only if excess tissue is available. Alternatively, protein can be harvested from in vitro studies using cell culture models (minimum 0.5×10^6 smooth muscle or endothelial cells).

4. To help designate the order of the samples, load more protein marker (5 µL) on the left side of the gel and less volume (2.5 µL) on the right side.

5. Cutting the right corner translates further designates the samples on the gel. The cut right side of the gel corresponds to the lesser volume of protein ladder loaded.

6. Cutting the right corner translates the order and designation of your samples on the gel to the blot. The cut right side of the gel corresponds to cut right side of the membrane.

7. Follow any standard Coomassie blue destain protocol to confirm proteins in the gel, and quantify if desired.

8. 300 mM Glycine-PBS should be filtered prior to use to remove residual glycine crystal and other debris that may interfere later with image acquisition.

9. Make a humidifying chamber by lining the bottom of a flat container with wet paper towels, such as an empty slide box with a lid. If the container does not have a lid, carefully cover with plastic wrap to prevent slides from drying during long incubations.

10. It is best to protect fluorescent secondary antibodies from direct light. Close the lid of the humidifying chamber or cover with foil while the sections incubate with antibodies.

11. Cover slides with foil during all washes after this point to protect sections bound with fluorescent secondary antibodies.

12. Actual H_2S concentration is taken as 30% of the NaHS concentration in the calculation (0, 2.3, 9.38, 18.75, 37.5, 75, 150, 300 µM).

Acknowledgments

The present study was supported in part by National Institutes of Health (NIH) grants RO1 HL70562, R21 HL98746, and RO3 HD84972 to D.B.C.

References

1. Wang R (2012) Physiological implications of hydrogen sulfide: a whiff exploration that blossomed. Physiol Rev 92:791–896

2. Gadalla MM, Snyder SH (2010) Hydrogen sulfide as a gasotransmitter. J Neurochem 113:14–26

3. Mustafa AK, Gadalla MM, Sen N, Kim S, Mu W, Gazi SK, Barrow RK, Yang G, Wang R, Snyder SH (2009) H2S signals through protein S-sulfhydration. Sci Signal 2: ra72

4. Leffler CW, Parfenova H, Basuroy S, Jaggar JH, Umstot ES, Fedinec AL (2011) Hydrogen sulfide and cerebral microvascular tone in newborn pigs. Am J Physiol Heart Circ Physiol 300:H440–H447

5. Bhatia M (2005) Hydrogen sulfide as a vasodilator. IUBMB Life 57:603–606

6. Shibuya N, Mikami Y, Kimura Y, Nagahara N, Kimura H (2009) Vascular endothelium expresses 3-mercaptopyruvate sulfurtransferase and produces hydrogen sulfide. J Biochem 146:623–626

7. Zhao W, Zhang J, Lu Y, Wang R (2001) The vasorelaxant effect of H(2)S as a novel endogenous gaseous K(ATP) channel opener. EMBO J 20:6008–6016

8. Li Y, Zang Y, Fu S, Zhang H, Gao L, Li J (2012) H2S relaxes vas deferens smooth muscle by modulating the large conductance Ca2+ −activated K+ (BKCa) channels via a redox mechanism. J Sex Med 9:2806–2813

9. Papapetropoulos A, Pyriochou A, Altaany Z, Yang G, Marazioti A, Zhou Z, Jeschke MG, Branski LK, Herndon DN, Wang R, Szabo C (2009) Hydrogen sulfide is an endogenous stimulator of angiogenesis. Proc Natl Acad Sci U S A 106:21972–21977

10. Yang G, Wu L, Jiang B, Yang W, Qi J, Cao K, Meng Q, Mustafa AK, Mu W, Zhang S, Snyder SH, Wang R (2008) H2S as a physiologic vasorelaxant: hypertension in mice with deletion of cystathionine gamma-lyase. Science 322:587–590

11. Magness RR, Rosenfeld CR (1986) Systemic and uterine responses to alpha-adrenergic stimulation in pregnant and nonpregnant ewes. Am J Obstet Gynecol 155:897–904

12. Magness RR, Rosenfeld CR (1989) Local and systemic estradiol-17 beta: effects on uterine and systemic vasodilation. Am J Phys 256: E536–E542

13. Rosenfeld CR, Chen C, Roy T, Liu X (2003) Estrogen selectively up-regulates eNOS and nNOS in reproductive arteries by transcriptional mechanisms. J Soc Gynecol Investig 10:205–215

14. Naden RP, Rosenfeld CR (1985) Systemic and uterine responsiveness to angiotensin II and norepinephrine in estrogen-treated nonpregnant sheep. Am J Obstet Gynecol 153:417–425

15. Killam AP, Rosenfeld CR, Battaglia FC, Makowski EL, Meschia G (1973) Effect of estrogens on the uterine blood flow of oophorectomized ewes. Am J Obstet Gynecol 115:1045–1052

16. Magness RR, Phernetton TM, Gibson TC, Chen DB (2005) Uterine blood flow responses to ICI 182 780 in ovariectomized oestradiol-17beta-treated, intact follicular and pregnant sheep. J Physiol 565:71–83

17. Ford SP (1982) Control of uterine and ovarian blood flow throughout the estrous cycle and pregnancy of ewes, sows and cows. J Anim Sci 55(Suppl 2):32–42

18. Gibson TC, Phernetton TM, Wiltbank MC, Magness RR (2004) Development and use of an ovarian synchronization model to study the effects of endogenous estrogen and nitric oxide on uterine blood flow during ovarian cycles in sheep. Biol Reprod 70:1886–1894

19. Nelson SH, Steinsland OS, Suresh MS, Lee NM (1998) Pregnancy augments nitric oxide-dependent dilator response to acetylcholine in the human uterine artery. Hum Reprod 13:1361–1367

20. Lang U, Baker RS, Braems G, Zygmunt M, Kunzel W, Clark KE (2003) Uterine blood flow—a determinant of fetal growth. Eur J Obstet Gynecol Reprod Biol 110(Suppl 1): S55–S61

21. Lechuga TJ, Zhang H, Sheibani L, Karim M, Jia J, Magness RR, Rosenfeld CR, Chen DB (2015) Estrogen replacement therapy in ovariectomized nonpregnant ewes stimulates uterine artery hydrogen sulfide biosynthesis by selectively upregulating cystathionine beta synthase expression. Endocrinology 156:2288–2298

22. O'Leary P, Boyne P, Flett P, Beilby J, James I (1991) Longitudinal assessment of changes in reproductive hormones during normal pregnancy. Clin Chem 37:667–672

23. Sheibani L, Lechuga TJ, Zhang H, Hameed A, Wing DA, Kumar S, Rosenfeld CR, Chen DB (2017) Augmented H2S production via cystathionine-beta-synthase upregulationplays a role in pregnancy-associated uterine vasodilation. Biol Reprod 96:664–672.

24. Chen DB, Westfall SD, Fong HW, Roberson MS, Davis JS (1998) Prostaglandin F2alpha stimulates the Raf/MEK1/mitogen-activated protein kinase signaling cascade in bovine luteal cells. Endocrinology 139:3876–3885

Chapter 4

Measurement of Protein Persulfidation: Improved Tag-Switch Method

Emilia Kouroussis, Bikash Adhikari, Jasmina Zivanovic, and Milos R. Filipovic

Abstract

Hydrogen sulfide (H_2S) is an endogenously produced signaling gasotransmitter, generated by the enzymes cystathionine γ-lyase, cystathionine β-synthase, and 3-mercaptopyruvate sulfurtransferase. The involvement of H_2S in numerous physiological, as well as pathophysiological conditions, was established over the past decade. However, the exact mechanism(s) of regulation of the biological functions by H_2S are under active investigations. It is proposed that the oxidative posttranslational modification of protein cysteine residues, known as persulfidation, could be the main mechanism of action of H_2S. Protein persulfides show similar reactivity to thiols, which represents one of the main obstacles in the development of a reliable method for detection of this specific protein modification. Subsequently, having a selective method for persulfide detection is of utmost importance in order to fully understand the physiological and pathophysiological role of H_2S. Several methods have been proposed for the detection of protein persulfidation, all of which are highlighted in this chapter. Furthermore, we provide a detailed description and protocol for the first selective persulfide labeling method, a tag-switch method, developed in our group.

Key words Hydrogen sulfide, Gasotransmitter, Oxidative posttranslational modification, Persulfide, Tag-switch assay, CN-BOT, CN-Cy3, MSBT

1 Introduction

1.1 Protein Persulfidation

Over the past decade, hydrogen sulfide (H_2S) has emerged as the third gasotransmitter alongside nitric oxide (NO) and carbon monoxide (CO) [1, 2]. The production of H_2S in the cell has been found to be controlled by at least three enzymes, cystathionine β-synthase (CBS), cystathionine γ-lyase (CSE), and 3 mercaptopyruvate sulfur transferase (MPST). These enzymes are expressed at different levels in different tissues and control H_2S production with different efficiencies. H_2S has been shown to completely or partially regulate various physiological and pathophysiological processes [3–8]. The main mechanism by which H_2S has been proposed to regulate biological functions is the formation

Jerzy Bełtowski (ed.), *Vascular Effects of Hydrogen Sulfide: Methods and Protocols*, Methods in Molecular Biology, vol. 2007, https://doi.org/10.1007/978-1-4939-9528-8_4, © Springer Science+Business Media, LLC, part of Springer Nature 2019

of persulfides on specific protein cysteine residues (referred to as S-sulfhydration or persulfidation), an oxidative posttranslational modification that could change the protein structure and activity [9].

Protein persulfidation can potentially explain the ample effects that H_2S has been documented to exhibit in the cell [10]. For example, parkin, a key Parkinson's disease (PD)-associated protein, is a documented case of a protein affected by oxidative/nitrosative stress [11–16]. Parkin functions as an E3 ubiquitin ligase, i.e., it catalyzes the thioester transfer of ubiquitin moieties to a variety of proteins. The loss of protein activity has been shown to be a reason for PD [17]. A recent study by Snyder's group demonstrated that parkin gets persulfidated, which consequently causes an increase of its activity [18]. This overactivation of parkin's function could rescue neurons from cell death by removing damaged proteins. The use of H_2S donors may therefore help in the early treatment of PD.

Despite the growing interest for protein persulfidation, there is still limited evidence in the literature regarding the exact mechanism(s) by which proteins are persulfidated. Initial studies of protein persulfidation were based on the incorrect assumption that the deprotonated form of free thiols (thiolates) could react directly with H_2S, resulting in the formation of persulfides. This reaction is, however, thermodynamically unfavorable and does not occur [19–21]. The exact mechanism by which proteins are modified by H_2S is an important question, which, when clearly understood, could be a crucial linker toward the unraveling of H_2S signaling. Several mechanisms have been proposed [21], but the most plausible is the reaction of H_2S with oxidized thiols, more precisely, disulfides or sulfenic acids.

The proposed reaction of H_2S with disulfides could represent a route for H_2S consumption in the extracellular matrix and plasma, for example, where higher levels of disulfides are present. However, recent studies suggested that the reaction of H_2S with disulfides is too slow to be of physiological relevance [22]. Interestingly, the reaction of H_2S with sulfenic acids (RSOH) has a rate constant higher than that of the reaction of other biological thiols with RSOH [22]. The treatment of cells with hydrogen peroxide (H_2O_2) showed increased intracellular levels of protein persulfidation, a process that could be completely suppressed by inhibiting endogenous H_2S production [22].

The pK_a of persulfides is lower than that of their corresponding thiols [21–23] suggesting that at physiological conditions, the majority of persulfides would be in their deprotonated form (R-S-S$^-$), making them "super" nucleophilic. This should dramatically increase persulfides' reactivity when compared to their corresponding thiols. Indeed the rate constant of the reaction of protein persulfides with peroxynitrite (powerful oxidant formed in the diffusion controlled reaction of superoxide with nitric oxide) is found to be one order of magnitude greater than for the reaction of peroxynitrite with its corresponding thiol [22].

All of these observations led to a suggestion that protein persulfidation could serve as a protection mechanism where, via protein persulfidation, the cellular milieu gets protected from irreversible protein hyper-oxidation, induced by a high amount of reactive oxygen (ROS) and nitrogen species (RNS) [21]. Namely, thiol oxidation, which initially starts with the formation of sulfenic acids (still reversible modification) could proceed further with the irreversible formation of sulfonic acids. H_2S could react with the sulfenic acids instead, preventing this oxidation. In addition, persulfidated protein will react faster with ROS/RNS and form an adduct that could be cleaved by the action of certain enzymes restoring the free thiol.

To exert a regulatory function similar to that of phosphorylation/dephosphorylation or S-nitrosation/denitrosation, S-persulfidation levels must be enzymatically modulated [24]. Intracellular protein disulfides and S-glutathionylation levels are controlled by the thioredoxin (Trx) system [25]. The enzymatic system, consisting of Trx, thioredoxin reductase (TrxR), and NADPH, represents the main disulfide reductase system in cells. In addition to its disulfide reductase activity Trx cleaves the persulfides one order of magnitude more efficiently than it reduces corresponding disulfides. The inhibition of the Trx system leads to an increase of intracellular persulfides, confirming that this process occurs in the cells as well. Significantly lower total sulfane sulfur levels were detected in HIV patients with high viral load (and high circulatory Trx levels) compared to the treated patients, which provides evidence that Trx acts as depersulfidase additionally in vivo [26].

Recent studies have shown that proteins such as NF-κB, Keap1, GAPDH, KATP-channels, PTP1B, etc. undergo protein persulfidation [2, 19, 27–30]. It has been estimated that up to 25% of proteins are persulfidated [31] making this modification almost as abundant as phosphorylation, and thus being of crucial importance for cells. However, due to the lack of an accurate and selective method for persulfide detection, the total amount of persulfidated protein remains questionable.

Persulfides are made of two sulfurs with different electronegativities, which results in it demonstrating two modes of reactivity. The sulfur covalently bound to carbon (RSSH) is considered a sulfane sulfur, with a formal charge of 0. Subsequently, this sulfane sulfur is susceptible to nucleophilic attack [32]. The persulfide's terminal sulfur (RSSH), however, possesses a formal charge of -1, which makes it susceptible to reaction with electrophiles [32]. As a consequence, one of the main obstacles in the development of a reliable and selective detection method for persulfides is the similar reactivity of persulfides to other sulfur species, especially thiolates.

According to literature, there has been a lot of debate on whether the detection methods used for persulfide labeling are indeed selective for protein persulfides [21]. This chapter provides an overview of the currently reported methods for protein persulfide labeling, with particular emphasis on the tag-switch method developed in our group.

1.2 Modified Biotin-Switch Method

The first method proposed in the literature for the labeling and detection of protein persulfides was by Mustafa et al. [31]. This method was a modification of the method originally used for the detection of protein S-nitrosation in proteins, known as the biotin-switch assay [33]. This assay for S-nitrosothiols was developed as a three-step method, where free thiols were initially blocked with the electrophilic alkylating agent, S-methyl methanethiosulfonate (MMTS), and after removing the excess of MMTS, ascorbate was added to reduce the S-nitrosothiols to free thiols. The released thiols were then selectively conjugated with N-[6-(biotinamido) hexyl]-3′-(2-pyridyldithio) propionamide (biotin-HPDP) and captured by streptavidin beads.

Mustafa and colleagues proposed a modified biotin-switch technique (modified-BST), illustrated in Fig. 1, in which protein persulfides were postulated to remain unreacted after the blocking of thiols with MMTS. Hence after the excess MMTS is removed, the free persulfides can be labeled with the use of biotin-HPDP, as shown in Fig. 1. Using this method, Mustafa et al. claimed that up to 25% of proteins in cell lysates are modified by H_2S, under basal conditions [31].

MMTS is extensively used for the detection of protein S-nitrosation and used in the in vivo trapping of the thiol-disulfide state of proteins [34, 35]. However, caution must be taken when using MMTS as Karala and Ruddock documented that MMTS generates artificial intermolecular and intramolecular protein disulfide bonds, which can give rise to the misinterpretation of results [36].

The chemical foundation of the modified biotin-switch technique was the selective reactivity of MMTS with thiols. However,

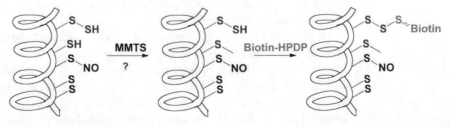

Fig. 1 Schematic overview of modified biotin-switch assay. The first step of this method was based on a chemically questionable premise that protein persulfides would not react with electrophilic thiol-blocking reagent S-methyl methanethiosulfonate (MMTS). In the subsequent step, persulfides are labeled with N-[6-(biotinamido) hexyl]-3′-(2′-pyridyldithio) propionamide (biotin-HPDP)

the potential nucleophilicity and hence reactivity of persulfides with the electrophilic MMTS was not investigated. Pan and Carroll [32] tested the reactivity of MMTS using low molecular weight (GSH persulfide) and protein persulfide models (papain persulfide and glutathione peroxidase 3 (Gpx3) persulfide). Their results demonstrated that the alkylated products were present in the product mixture following the reaction with MMTS. In the case of the low MW model, GSH persulfide, the alkylated product was obtained as a minor product, while with Gpx3 persulfide and papain persulfide, the alkylated product was obtained as the major product. Additionally, the reactivity of these persulfide models toward electrophilic and nucleophilic reagents was tested, giving a further insight into persulfide reactivity. The nucleophilicity of the terminal sulfur of tested persulfides (RSSH) was reaffirmed, showing without a doubt that they react with MMTS (and its brominated analogue BBMTS) as readily as free thiols.

1.3 Cy5-Maleimide Labeling and Further Modifications of the Method

Snyder's group also [27] proposed a modified NEM (N-ethylmaleimide) method for the persulfide labeling of purified proteins (Fig. 2). Cy-5 labeled maleimide was used as a thiol-blocking reagent, to block both the persulfides and free thiol of tested protein sample. The product of Cy5-maleimide and persulfide is actually a disulfide that can be cleaved by dithiothreitol (DTT). The samples were then treated with DTT and the decrease of in-gel fluorescent signal monitored as readout for the persulfide levels (Fig. 2). Simplicity of this method, as well as commercial availability of the reagents, represents the two main advantages; however the analysis of complex protein mixtures becomes more difficult.

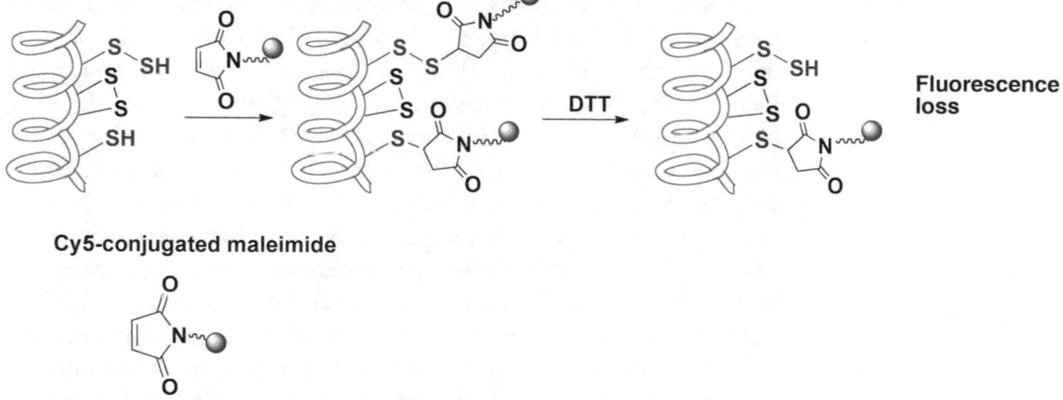

Fig. 2 Schematic overview of Cy5-maleimide method. In this method both persulfide and free thiol would be blocked by the thiol fluorescently labeled N-ethyl maleimide (Cy5-conjugated maleimide). The adduct of persulfide and Cy5-maleimide is a disulfide that will be then cleaved by the DTT leading to a decrease of the in-gel fluorescence signal in the samples containing persulfides

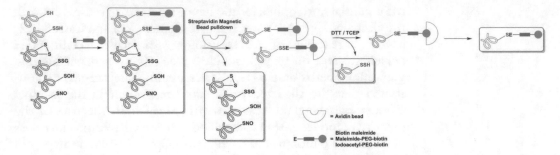

Fig. 3 Chemical modifications of method proposed by Sen et al. [27]. Proteins get initially labeled with alkylating agent such as biotin-maleimide, maleimide-PEG-biotin, or iodoacetyl-PEG-biotin which label both thiols and persulfides. Labeled proteins then get pulled down by streptavidin (or avidin) beads, cleaned from the rest of the mixture, and persulfidated proteins get eluted by some reducing reagent such as DTT or TCEP. Cuevasanta et al. [22], Gao et al. [37], and Longen et al. [39] trypsinized the labeled protein first and then pulled down the peptides with streptavidin beads, while Dóka et al. [38] worked with intact proteins

We described a slight modification of the method proposed by Snyder's group [22], which has since been used by several other authors with minor modifications (Fig. 3). In this methodological approach, free thiols and persulfides are initially labeled with biotinylated maleimide. Labeled proteins are then digested using trypsin and biotinylated peptides separated using streptavidin agarose beads. The alkylation of persulfides leads to the formation of a disulfide, making the elution of persulfidated peptides from streptavidin beads by DTT rather easy. After centrifugation, the eluant was analyzed by LC-MS/MS. Gao et al. [37] used a similar method to detect persulfides in cell lysates. For the blocking step, an alternative reagent was used, maleimide-PEG_2-biotin (NM-biotin), which was followed by the binding of the biotin-labeled proteins on an avidin column.

Other thiol-blocking reagents could be used instead of maleimides, such as Iodoacetyl-PEG_2-Biotin (IAB). Some authors tried to name this method, calling it ProPerDP [38] or qPerS-SID [39]. In the ProPerDP method, instead of digesting the alkylated proteins prior to streptavidin separation, Dóka et al. [38] separated the whole proteins on streptavidin beads, eluting the persulfidated proteins with TCEP (tris(2-carboxyethyl)phosphine). Separating the whole proteins by this approach is more prone to artifacts. In addition to inevitable elution of proteins connected by inter- and intramolecular disulfides and which do not necessarily have to contain any persulfides, the actual yield of eluted persulfidated proteins is underestimated. For example, if a protein contains two cysteine residues, of which only one is persulfidated, then the chances for the persulfidated protein to be eluted from streptavidin beads are 50%. For example, in case of another protein TRPA1 [40] whose persulfidation is postulated to contains 21 cysteine residues, those chances would be 1/21. Therefore, it is not surprising that

Dóka et al. [38] reported very low protein persulfidation using the ProPerDP approach.

In the qPerS-SID method, Longen and colleagues [39] used this approach for quantitative proteomic analysis of protein persulfidation, where control cells were grown on standard cell medium, while the cells treated with H_2S donors were grown on SILAC (stable isotope labeling with amino acid in culture) medium. The cells were treated with iodoacetyl-PEG2-biotin (IAB) to block thiols and persulfides, samples were mixed in a ratio of 1:1 and digested by trypsin. Peptides containing cysteine and persulfide were separated from other peptides in the mixture using streptavidin beads. Bound persulfidated peptides were separated from the bound peptides with cysteine using the reducing agent (TCEP) to cleave the disulfide bond in the persulfides. In the following step, peptides were treated with iodoacetamide (IAM) to improve their detection by LC-MS/MS analysis. However, certain limitations arise concerning this method. The critical step of this method, similar to that of the ProPerDP method [38], is its reduction step and subsequent breaking of the disulfide bonds, hence not being selective only to persulfides. As a result, the intramolecular disulfide bonds will also be reduced causing false-positive results. Furthermore, following the reduction, the use of IAM as a thiol-blocking reagent can additionally label primary amines [41] and sulfenic acids [42], as the authors suggested. Indeed, the authors did not see any significant increase of protein persulfidation in the cells treated with the most used H_2S donors, NaHS or Na_2S, contrary to all other published studies, which further questions this methodological approach.

To determine the persulfidation of protein tyrosine phosphatase 1B (PTP1B) in the cell lysate, Krishnan and colleagues [28] used iodoacetic acid (IAA) as a thiol-blocking reagent (Fig. 4). Free thiols and persulfides will be blocked since the persulfide reactivity is similar to that of the free thiols. However, in the second step, they used DTT to cleave the alkylated persulfides in order to form free thiols. Next, the free thiols were labeled with iodoacetamide-linked

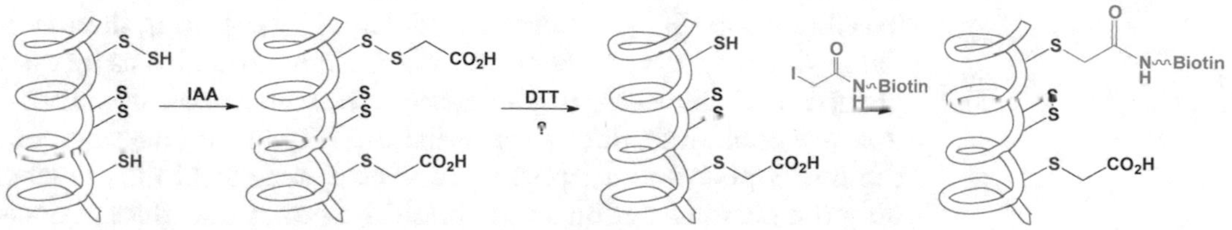

Fig. 4 Schematic overview of persulfide labeling approach proposed by Krishnan et al. [28]. In this method, iodoacetic acid (IAA) is used to initially block both free thiols and protein persulfides. In the subsequent steps, alkylated persulfide is cleaved with DTT and then labeled with iodoacetamide-linked biotin (IAP). Although DTT would indeed cleave this adduct, it is unclear how this method distinguishes the persulfides from intra- and intermolecular disulfides and S-nitrosothiols, which would also be reduced by DTT

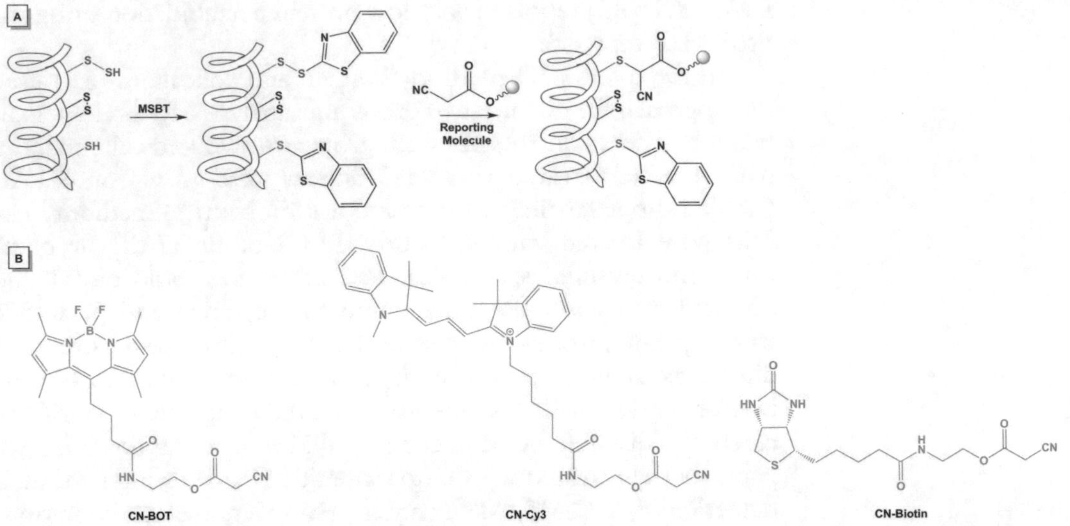

Fig. 5 Tag-switch method for persulfide labeling. (**a**) Schematic overview of labeling steps in tag-switch method. Methylsulfonyl benzothiazole (MSBT) is used to block thiols and persulfides in the first step, followed by the tag switch with cyanoacetate derivatives that carry a reporting molecule, in the second step. (**b**) Structures of reporting molecules used to label protein persulfides from the cells

biotin (IAP) and purified on streptavidin beads. The main concern with this method is the use of DTT as a reducing reagent in the second step of the protocol. Namely, DTT would cleave the alkylated persulfides, but it would also cleave all the disulfide bonds in the protein and consequently cause false-positive results.

1.4 Tag-Switch Method

We proposed that persulfidation can be selectively detected by the tag-switch method (i.e., using two reagents to label protein persulfides in two steps) [10, 19]. In the first step, a thiol-blocking reagent should be introduced and tagging both P-SH and P-SSH to form an intermediate product (Fig. 5a). If an appropriate thiol blocking reagent is employed, the disulfide bonds in the persulfide adducts may show enhanced reactivity to certain nucleophiles than common disulfides in proteins. We screened a series of carbon-based nucleophiles as potential candidates [19]. Among these candidates, methyl cyanoacetate was particularly attractive as its ester group could allow easy installation of reporting molecules. Therefore, we could introduce a tag-switching reagent (containing both the nucleophile and a reporting molecule, such as biotin) to label only the persulfide adducts. It should be noted that thiol adducts from the first step are thioethers, which are not expected to react with the nucleophile. Moreover, in contrast to previous methods, even if the free thiol is not completely blocked, we should not expect any misidentification of persulfidated proteins because the tag-switch reagent is a nucleophile, not an electrophile.

A major challenge in this technology was whether the newly generated disulfide linkages from persulfide moieties could display a unique reactivity to a suitable nucleophile to an extent that it is differentiated from common protein disulfides. We envisioned that a reagent, which would give a mixed aromatic disulfide linkage when reacting with persulfides (-S-SH), could exert the reactivity criteria for our tag-switch technology (Fig. 5a). Indeed, by combining methylsulfonyl benzothiazole (MSBT-A) as a thiol-blocking reagent in the first step, and a biotinylated derivative of methyl cyanoacetate as a nucleophile in the second step, we could efficiently label protein persulfides (Fig. 5a) [19]. To test the selectivity of this method, we produced P-SSH on bovine serum albumin (BSA) and compared its reactivity with the tag-switch assay using glutathionylated, sulfenylated, and normal BSA (which by definition contains both intramolecular disulfides and one reactive cysteine). Only P-SSH was labeled and could be pulled down by streptavidin beads, suggesting the applicability of tag-switch assay for wide proteomic analysis [43].

The original assay used was with a biotinylated cyanoacetic acid tag, which required Western blot transfer and streptavidin or antibodies for visualization [10, 19]. To increase sensitivity, we synthesized two new cyanoacetic acid derivatives, with a fluorescent BODIPY moiety (CN-BOT) or a Cy3-dye (CN-Cy3) (Fig. 5b) [26]. Both new tags labeled persulfidated human serum albumin (HSA-SSH) resulting in the formation of fluorescent products. CN-BOT was used for the labeling of cells for microscopy (Fig. 6) and CN-Cy3 for the labeling of cell lysates. This is because the former showed low fluorescence after gel fixation, while the latter proved to be very difficult to wash out from fixed cells. This lead to the discovery that the thioredoxin/thioredoxin reductase system is essentially involved in the removal of protein persulfidation, thus acting as protein depersulfidase system. Using this improved tag-switch assay, we also demonstrated the role of protein persulfidation in a *D. melanogaster* disease model of spinocerebellar ataxia type 3 (SCA3) [44].

2   Materials

2.1   "Improved Tag-Switch" Assay for In-gel Detection

1. Ham's F12: DMEM (1:1) medium supplemented with 2 mM glutamine, 1% nonessential amino acids, and 10% fetal bovine serum.

2. PBS.

3. HEN buffer: 50 mM HEPES, 0.1 mM EDTA, 1.5% SDS, 1% NP-40, 1% protease inhibitor cocktail, and 10 mM MSBT pH 7.4. *See* **Note 1**.

4. Methanol.

CN-BOT	DAPI	Merge

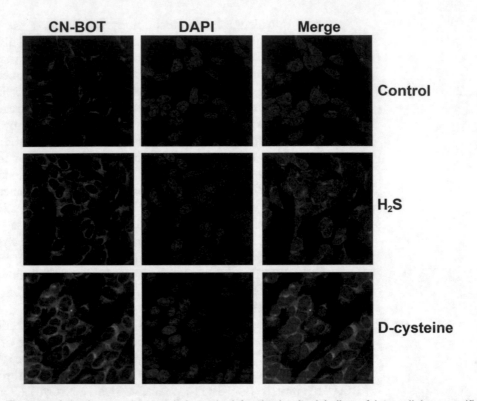

Fig. 6 The use of the improved tag-switch method for the in situ labeling of intracellular persulfides in SH-SY5Y neuroblastoma cells. Cells were treated with 100 μM Na$_2$S or 2 mM D-cysteine (substrate for 3-mercaptopyruvate sulfur transferase, MPST) for 1 h, to increase the levels of intracellular persulfidation. Labeling was performed as described in the protocol, with the green fluorescence originating from CN-BOT and blue from the use of DAPI to stain the nuclei. An obvious increase of intracellular persulfidation was achieved with exogenous treatment with Na$_2$S and even stronger effect was visible in the cells treated with 2 mM D-cysteine

5. Chloroform.

6. 50 mM HEPES with 3% SDS (pH 7.4).

7. Neocuproine hydrate.

8. 2.5 mM CN-Cy3 in acetonitrile.

2.2 "Improved Tag-Switch" Assay for In Situ Detection of Intracellular Persulfide

1. PBS.

2. Methanol.

3. Acetone.

4. 50 mM HEPES buffer (pH 7.4).

5. 10 mM MSBT dissolved into 70% 50 mM HEPES (pH 7.4)/ 30% methanol.

6. Triton X-100.

7. 2.5 mM CN-BOT in acetonitrile.

8. DAPI.

3   Methods

3.1   "Improved Tag-Switch" Assay

1. Grow SH-SY5Y cells in Ham's F12: DMEM medium, supplemented with 2 mM glutamine, 1% non-essential amino acids, and 10% fetal bovine serum at 37 °C and 5% CO_2 in T-75 cell culture flasks.

2. Treat the cells with respective compounds over 1 h.

3. Wash the cells twice with warm sterile PBS.

4. Lyse the cells by adding 800 μL HEN buffer that contains 10 mM MSBT to the T-75 flasks.

5. Incubate the cells on ice for 10 min with occasional scrapping of the flask surface with a cell scrapper.

6. Transfer the lysate to tubes and incubate for 1 h at 37 °C.

7. Precipitate the proteins from the lysate by chloroform/methanol precipitation. Start by adding methanol (MeOH) first (1/1, v/v) followed by vigorous vortexing and then add chloroform (final: $4/4/1$, water/MeOH/CHCl$_3$). Vortex the sample and centrifuge ($20{,}000 \times g$, 20 min, 4 °C). Proteins will form a visible intermediate layer pellet between the chloroform and MeOH/water fraction. To wash the protein pellet, remove the upper fraction, and replace with MeOH, mix, and centrifuge again ($20{,}000 \times g$, 20 min, 4 °C).

8. Dry the precipitated pellet and resuspend in 300 μL 50 mM HEPES with 3% SDS.

9. Incubate the protein solutions with 60 μM CN-Cy3 (by adding 3 μL of a 25 mM stock solution) for 1 h at 37 °C. *See* **Note 2**.

10. Resolve by SDS-PAGE under non-reducing conditions, and record the gels on a Cy3 channel.

3.2   "Improved Tag-Switch" Assay for In Situ Detection of Intracellular Persulfide

1. Grow cells in μ-dishes (35 mm, high) obtained from Ibidi® (Martinsried, Germany), following manufacturer's instructions.

2. Treat the cells with respective compounds over 1 h.

3. After treatments, wash the cells twice with warm sterile PBS.

4. Fix the cells by incubation with ice-cold methanol at −30 °C for 20 min. Remove methanol, and add ice-cold acetone to the cells for 5 min at −30 °C. Wash the dishes with PBS buffer.

5. Incubate the cells with 0.5 mL MSBT/HEPES/methanol solution at room temperature overnight.

6. Wash the cells five times with PBS, and incubate with 0.5 mL of 25 μM CN-BOT (obtained by adding 5 μL of CN-BOT stock solution) in HEPES at 37 °C for 1 h.

7. Wash the cells five times with PBS and stain with DAPI, following manufacturer's instruction.

8. Wash the cells again and visualize using an LSM 780 confocal laser scanning system (Carl Zeiss MicroImaging).

4 Notes

1. MSBT shows poor solubility. Phosphate buffers further decrease its solubility leading to the precipitation. Stock solutions should be prepared in methanol and added into HEN buffer which already contained 1.5% SDS, 1% NP-40, and 1% protease inhibitor cocktail.

2. Sulfenic acids could react with cyanoacetic acid-derived probes. Although the chances that they remain intact in the cells after the first five steps are minor (indeed we confirm that they do not interfere with the labeling [19]). An additional step could be introduced immediately after the fixation where the cells would be incubated with 1 mM dimedone in 70% 50 mM HEPES (pH 7.4)/30% methanol for 1 h at 37 °C, washed carefully, and then continue labeling as described in **step 5**.

Acknowledgments

Authors would like to thank CNRS/INSERM Atip-Avenir and the "Investments for the future" Programme IdEx Bordeaux (ANR-10-IDEX-03-02) for their financial support.

References

1. Wang R (2002) Two's company, three's a crowd: can H2S be the third endogenous gaseous transmitter? FASEB J 16:1792–1798. https://doi.org/10.1096/fj.02-0211hyp

2. Mustafa AK, Gadalla MM, Sen N et al (2009) H2S signals through protein S-sulfhydration. Sci Signal 2:ra72. https://doi.org/10.1126/scisignal.2000464

3. Yang G, Wu L, Jiang B et al (2008) H2S as a physiologic vasorelaxant: hypertension in mice with deletion of cystathionine gamma-lyase. Science 322:587–590. https://doi.org/10.1126/science.1162667

4. Kimura H, Nagai Y, Umemura K, Kimura Y (2005) Physiological roles of hydrogen sulfide: synaptic modulation, neuroprotection, and smooth muscle relaxation. Antioxid Redox Signal 7:795–803. https://doi.org/10.1089/ars.2005.7.795

5. Papapetropoulos A, Pyriochou A, Altaany Z et al (2009) Hydrogen sulfide is an endogenous stimulator of angiogenesis. Proc Natl Acad Sci U S A 106:21972–21977. https://doi.org/10.1073/pnas.0908047106

6. Li L, Bhatia M, Moore PK (2006) Hydrogen sulphide—a novel mediator of inflammation? Curr Opin Pharmacol 6:125–129. https://doi.org/10.1016/j.coph.2005.10.007

7. Paul BD, Snyder SH (2012) H2S signalling through protein sulfhydration and beyond. Nat Rev Mol Cell Biol 13:499–507. https://doi.org/10.1038/nrm3391

8. Kabil O, Motl N, Banerjee R (2014) H2S and its role in redox signaling. Biochim Biophys

Acta 1844:1355–1366. https://doi.org/10.1016/j.bbapap.2014.01.002

9. Filipovic MR, Zivanovic J, Alvarez B, Banerjee R (2017) Chemical biology of H S signaling through persulfidation. Chemical Reviews 118(3):1253–1337

10. Park CM, Macinkovic I, Filipovic MR, Xian M (2015) Use of the "Tag-Switch" method for the detection of protein S-sulfhydration, 1st edn. Methods Enzymol. https://doi.org/10.1016/bs.mie.2014.11.033

11. Meng F, Yao D, Shi Y et al (2011) Oxidation of the cysteine-rich regions of parkin perturbs its E3 ligase activity and contributes to protein aggregation. Mol Neurodegener 6:34. https://doi.org/10.1186/1750-1326-6-34

12. Bossy-Wetzel E, Schwarzenbacher R, Lipton SA (2004) Molecular pathways to neurodegeneration. Nat Med 10:S2–S9. https://doi.org/10.1038/nm1067

13. Gu Z, Nakamura T, Yao D et al (2005) Nitrosative and oxidative stress links dysfunctional ubiquitination to Parkinson's disease. Cell Death Differ 12:1202–1204. https://doi.org/10.1038/sj.cdd.4401705

14. Cho D-H, Nakamura T, Fang J et al (2009) S-nitrosylation of Drp1 mediates β-amyloid-related mitochondrial fission and neuronal injury. Science 324:102–105. https://doi.org/10.1126/science.1171091

15. Fang J, Nakamura T, Cho D-H et al (2007) S-nitrosylation of peroxiredoxin 2 promotes oxidative stress-induced neuronal cell death in Parkinson's disease. Proc Natl Acad Sci 104:18742–18747. https://doi.org/10.1073/pnas.0705904104

16. Giasson BI, Lee VM-Y, Chung KK et al (2003) Are ubiquitination pathways central to Parkinson's disease? Cell 114:1–8. https://doi.org/10.1016/S0092-8674(03)00509-9

17. Dawson TM, Dawson VL (2010) The role of parkin in familial and sporadic Parkinson's disease. Mov Disord 25(Suppl 1):S32–S39. https://doi.org/10.1002/mds.22798

18. Vandiver MS, Paul BD, Xu R et al (2013) Sulfhydration mediates neuroprotective actions of parkin. Nat Commun 4:1626. https://doi.org/10.1038/ncomms2623

19. Zhang D, Macinkovic I, Devarie-Baez NO et al (2014) Detection of protein S-sulfhydration by a Tag-Switch technique. Angew Chem Int Ed 53:575–581. https://doi.org/10.1002/anie.201305876

20. Wedmann R, Bertlein S, Macinkovic I et al (2014) Working with "H2S": facts and apparent artifacts. Nitric Oxide 41:85–96. https://doi.org/10.1016/j.niox.2014.06.003

21. Filipovic MR (2015) Persulfidation (S-sulfhydration) and H2S. Handb Exp Pharmacol 230:29–59

22. Cuevasanta E, Lange M, Bonanata J et al (2015) Reaction of hydrogen sulfide with disulfide and sulfenic acid to form the strongly nucleophilic persulfide. J Biol Chem 290:26866–26880. https://doi.org/10.1074/jbc.M115.672816

23. Everett SA, Folkes LK, Wardman P, Asmus KD (1994) Free-radical repair by a novel perthiol: reversible hydrogen transfer and perthiyl radical formation. Free Radic Res 20:387–400

24. Lu J, Holmgren A (2014) The thioredoxin superfamily in oxidative protein folding. Antioxid Redox Signal 21:457–470. https://doi.org/10.1089/ars.2014.5849

25. Burke-Gaffney A, Callister MEJ, Nakamura H (2005) Thioredoxin: friend or foe in human disease? Trends Pharmacol Sci 26:398–404. https://doi.org/10.1016/j.tips.2005.06.005

26. Wedmann R, Onderka C, Wei S et al (2016) Improved tag-switch method reveals that thioredoxin acts as depersulfidase and controls the intracellular levels of protein persulfidation. Chem Sci. https://doi.org/10.1039/C5SC04818D

27. Sen N, Paul BD, Gadalla MM et al (2012) Hydrogen sulfide-linked sulfhydration of NF-κB mediates its antiapoptotic actions. Mol Cell 45:13–24. https://doi.org/10.1016/j.molcel.2011.10.021

28. Krishnan N, Fu C, Pappin DJ, Tonks NK (2011) H2S-Induced sulfhydration of the phosphatase PTP1B and its role in the endoplasmic reticulum stress response. Sci Signal 4:1–26. https://doi.org/10.1126/scisignal.2002329

29. Yang G, Zhao K, Ju Y et al (2013) Hydrogen sulfide protects against cellular senescence via S-sulfhydration of Keap1 and activation of Nrf2. Antioxid Redox Signal 18:1906–1919. https://doi.org/10.1089/ars.2012.4645

30. Hourihan JM, Kenna JG, Hayes JD (2013) The gasotransmitter hydrogen sulfide induces Nrf2-target genes by inactivating the Keap1 ubiquitin ligase substrate adaptor through formation of a disulfide bond between Cys-226 and Cys-613. Antioxid Redox Signal 19:465–481. https://doi.org/10.1089/ars.2012.4944

31. Mustafa AK, Gadalla MM, Snyder SH (2009) Signaling by gasotransmitters. Sci Signal 2:re2. https://doi.org/10.1126/scisignal.268re2

32. Pan J, Carroll KS (2013) Persulfide reactivity in the detection of protein S-sulfhydration. ACS

Chem Biol 8:1110–1116. https://doi.org/10.1021/cb4001052

33. Jaffrey SR, Erdjument-Bromage H, Ferris CD et al (2001) Protein S-nitrosylation: a physiological signal for neuronal nitric oxide. Nat Cell Biol 3:193–197. https://doi.org/10.1038/35055104

34. Daly TJ, Olson JS, Matthews KS (1986) Formation of mixed disulfide adducts at cysteine-281 of the lactose repressor protein affects operator and inducer binding parameters. Biochemistry 25:5468–5474

35. Peaper DR, Wearsch PA, Cresswell P (2005) Tapasin and ERp57 form a stable disulfide-linked dimer within the MHC class I peptide-loading complex. EMBO J 24:3613–3623. https://doi.org/10.1038/sj.emboj.7600814

36. Karala A, Ruddock LW (2007) Does S-methyl methanethiosulfonate trap the thiol-disulfide state of proteins? Antioxid Redox Signal 9:527–531. https://doi.org/10.1089/ars.2006.1473

37. Gao XH, Krokowski D, Guan BJ et al (2015) Quantitative H2S-mediated protein sulfhydration reveals metabolic reprogramming during the integrated stress response. eLife. https://doi.org/10.7554/eLife.10067

38. Dóka É, Pader I, Bíró A et al (2016) A novel persulfide detection method reveals protein persulfide- and polysulfide-reducing functions of thioredoxin and glutathione systems. Sci

Adv 2:e1500968. https://doi.org/10.1126/sciadv.1500968

39. Longen S, Richter F, Köhler Y et al (2016) Quantitative persulfide site identification (qPerS-SID) reveals protein targets of H2S releasing donors in mammalian cells. Sci Rep 6:29808. https://doi.org/10.1038/srep29808

40. Mishanina TV, Libiad M, Banerjee R (2015) Biogenesis of reactive sulfur species for signaling by hydrogen sulfide oxidation pathways. Nat Chem Biol 11:457–464. https://doi.org/10.1038/nchembio.1834

41. Boja ES, Fales HM (2001) Overalkylation of a protein digest with iodoacetamide. Anal Chem 73:3576–3582

42. Reisz JA, Bechtold E, King SB et al (2013) Thiol-blocking electrophiles interfere with labeling and detection of protein sulfenic acids. FEBS J 280:6150–6161. https://doi.org/10.1111/febs.12535

43. Ida T, Sawa T, Ihara H et al (2014) Reactive cysteine persulfides and S-polythiolation regulate oxidative stress and redox signaling. Proc Natl Acad Sci USA 111:7606–7611. https://doi.org/10.1073/pnas.1321232111

44. Snijder PM, Baratashvili M, Grzeschik NA et al (2015) Overexpression of cystathionine γ-lyase suppresses detrimental effects of spinocerebellar ataxia type 3. Mol Med 21:758. https://doi.org/10.2119/molmed.2015.00221

Chapter 5

ProPerDP: A *Protein Per*sulfide *D*etection *P*rotocol

Éva Dóka, Elias S. J. Arnér, Edward E. Schmidt, and Péter Nagy

Abstract

Persulfide or polysulfide formation on Cys residues is emerging as an abundant protein posttranslational modification, with important regulatory functions. However, as many other Cys oxidative modifications, per- and polysulfides are relatively labile, dynamically interchanging species, which makes their intracellular detections challenging. Here we report our recently developed highly selective method, *Protein Per*sulfide *D*etection *P*rotocol (ProPerDP), which can detect protein per- and polysulfide species in isolated protein systems, in blood plasma, or in cells and tissue samples. The method is easy to use and relatively inexpensive and requires only readily commercially available reagents. The biggest advantage of ProPerDP compared to other previously published persulfide detecting methods is the fact that in this protocol, all thiol and persulfide species are appropriately alkylated before any cell lysis step. This greatly reduces the potential of detecting lysis-induced oxidation-driven artifact persulfide formation.

Key words Protein persulfide, ProPerDP, Detection method, Biotin pulldown assay, Selective reduction

Abbreviations

A549	Adenocarcinomic human alveolar basal epithelial cells
ACN	Acetonitrile
BCA	Bicinchoninic acid
BCIP	5-Bromo-4-chloro-3′-indolyl phosphate p-toluidine salt
BSA	Bovine serum albumin
CHAPS	3-[(3-Cholamidopropyl)dimethylammonio]-1-propanesulfonate hydrate
Cys	Cysteine
DMEM-F12	Dulbecco's modified eagle medium with F12 nutrient mixture
DTNB	5,5′-Dithiobis(2-nitrobenzoic acid), Ellman's reagent
DTPA	Diethylenetriaminepentaacetic acid
DTT	Dithiothreitol
EDTA	Ethylenediaminetetraacetic acid
EGTA	Ethylene glycol-bis(β-aminoethyl ether)-N,N,N',N'-tetraacetic acid
EMEM	Eagle's minimum essential medium
FBS	Heat-inactivated fetal bovine serum
HBSS	Hank's Balanced Salt Solution
HEK293	Human embryonic kidney cells 293

Jerzy Bełtowski (ed.), *Vascular Effects of Hydrogen Sulfide: Methods and Protocols*, Methods in Molecular Biology, vol. 2007,
https://doi.org/10.1007/978-1-4939-9528-8_5, © Springer Science+Business Media, LLC, part of Springer Nature 2019

HEPES	4-(2-Hydroxyethyl)-1-piperazineethanesulfonic acid
HSA	Human serum albumin
IAB	EZ-Link™ Iodoacetyl-PEG2-Biotin
IAF	5-Iodoacetamido fluorescein
IAM	Iodoacetamide
NBT	Nitro-blue tetrazolium chloride
PBS	Phosphate-buffered saline
PIC	Protease inhibitor cocktail
Pipes	Piperazine-N,N'-bis(2-ethanesulfonic acid)
ProPerDP	Protein persulfide detection protocol
PVDF	Polyvinylidene fluoride
SB	SDS sample buffer, nonreducing, 4×
TCEP	Tris(2-carboxyethyl)phosphine
TE	100 mM Tris–HCl, 2 mM EDTA, pH = 7.4
TNB-	2-Nitro-5-thiobenzoate
TR/GR-null	Mouse liver lacking thioredoxin reductase and glutathione reductase
Trx	Thioredoxin
TTBS	20 mM Tris, 0.5 M NaCl, pH 7.5 + 0.05% Tween 20

1   Introduction

Per- and polysulfide formation refers to the chemical addition of one or more sulfur atoms to an acceptor sulfur atom and is emerging as a common and abundant modification on protein Cys residues [1, 2], which can be introduced by posttranslational [3–6] mechanisms. Due to the novel chemical features of per-/polysulfides compared to Cys thiol species [7–10], per-/polysulfides are widely proposed to exhibit Cys protecting from oxidative stress [6, 10] and regulatory functions [11, 12]. This chapter aims to give a detailed description of the *Protein Persulfide Detection Protocol* (ProPerDP) method, a recently published experimental procedure for the detection of protein persulfidation in biological samples including isolated proteins, cells, and tissues [1]. The technique is based on alkylation and biotin-streptavidin affinity pulldown of protein thiol and per-/polysulfide species followed by selective reduction of persulfidated proteins from the surface of the streptavidin-coated beads (Fig. 1).

The method was validated using human serum albumin (HSA) as a model protein with a single free cysteine residue and inorganic polysulfides as protein Cys per-/polysulfidating agents. The method was shown to be able to detect inorganic polysulfide-mediated HSA persulfidation in human blood plasma too, in a polysulfide concentration-dependent manner. It was also proved effective in crude cell lysates representing more complex protein mixtures. However, the most significant improvement factor of the method is its capability to detect the persulfide pool from intact

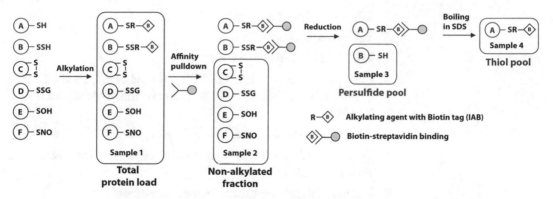

Fig. 1 Schematic workflow of the ProPerDP method (adapted from [1] with modifications, with permission from AAAS). The initial step of the protocol is the selective alkylation of thiol and persulfide moieties by EZ-Link Iodoacetyl-PEG2-Biotin (IAB) to generate the respective thioether and dialkyl disulfide products. The alkylated proteins are isolated from the original sample (Sample 1) by affinity purification using streptavidin-coated magnetic beads. Other modifications of Cys residues, which remain unaffected by alkylation, will be found in the supernatant after the pulldown step (Sample 2). In the key step of the protocol, the beads are incubated in reducing buffer where disulfide bonds originating from the persulfide groups are selectively reduced, and after separation, the supernatant can be analyzed with standard protein purification and identification methods as the persulfide fraction of the original sample (Sample 3). The biotin-streptavidin interaction at the surface of the beads can be reversed by boiling in SDS sample buffer, which will provide the free thiol pool of the original sample (Sample 4). The inset represents the biotin-streptavidin interaction, the diamond character (◇) refers to the biotin tag, and –R denotes the electrophile moiety of the alkylating agent (acetamide conjugated with PEG2 linker with iodide as the leaving group). The order and denotions of the samples are consistently used throughout this paper, with the respective samples referred to as *S1–S4*

cells without prior lysis or additional permeabilizing agents. The ProPerDP was further developed to detect protein persulfides in mouse liver tissue samples. Using ProPerDP on genetically modified cells and mouse liver tissue samples, we previously confirmed pivotal roles for the thioredoxin and the glutathione systems in the reduction of protein per-/polysulfide species and in the maintenance of in vivo persulfide homeostasis [1]. Another study suggested that the reactivity of Trx toward hydropersulfides is an order of magnitude higher compared to those toward the equivalent disulfide compounds [13].

Following the order from validation of the method on single purified protein through persulfide detection in diverse biological materials with different levels of complexity, detailed experimental conditions are provided in the present chapter for each application, presented as follows.

2 Materials

2.1 Buffers and Media

1. TE: 100 mM Tris–HCl pH = 7.4, using an appropriate transition metal chelating agent (e.g., 200 μM DTPA or 2 mM EDTA).

2. 100 mM phosphate buffer, pH 7.4.

3. PBS: phosphate-buffered saline.

4. HBSS: Hank's Balanced Salt Solution.

5. Cell lysis buffer: 40 mM HEPES, 50 mM NaCl, 1 mM EDTA, 1 mM EGTA, pH 7.4 + 1% protease inhibitor cocktail +1% CHAPS.

6. TTBS: 20 mM Tris, 0.5 M NaCl, pH 7.5 + 0.05% Tween 20.

7. DMEM-F12: Dulbecco's modified eagle medium with F12 nutrient mixture.

8. EMEM: Eagle's minimum essential medium.

9. FBS: heat-inactivated fetal bovine serum.

10. Penicillin.

11. Streptomycin.

12. Trypsin-EDTA solution for cell culture.

13. Cell counting dye, e.g., trypan blue.

14. Perfusion buffer: 150 mM NaCl, 100 mM pipes (pH 7.1).

15. Bradford assay reagent.

16. BCA assay kit.

2.2 Preparation of Polysulfide Reagent Solutions

1. $Na_2S \cdot 9H_2O$ (Sigma-Aldrich).

2. NaOCl solution (bleach).

3. DTNB: 5,5′-dithiobis(2-nitrobenzoic acid), Ellman's reagent.

4. Na_2S_3.

2.3 Samples

1. Human serum albumin (HSA, Sigma-Aldrich).

2. Human blood plasma.

3. A549 cells (ECACC).

4. HEK293 cells (ATCC).

5. Frozen tissue samples pre-perfused with IAB.

2.4 ProPerDP Method

1. Alkylating agent: IAB (EZ-Link™ Iodoacetyl-PEG2-Biotin, Thermo Fisher Scientific).

2. Microparticles, magnetic, streptavidin coated (Sigma-Aldrich).

3. Dynabeads® M-280 Streptavidin (Thermo Fisher Scientific).

4. TCEP: Tris(2-carboxyethyl)phosphine hydrochloride.

5. DTT: dithiothreitol.

6. SB: SDS sample buffer, nonreducing, 4×.

2.5 Western Blotting and Staining

1. Colloidal Coomassie stain.

2. PVDF membrane.

3. Blocking agent: 3% (w/v) BSA solution in TTBS.

4. Primary antibody: anti-albumin antibody produced in rabbit IgG fraction of antiserum (Sigma-Aldrich), diluted 1:10,000 in 3% BSA solution.

5. Secondary antibody: anti-rabbit IgG (whole molecule)-alkaline phosphatase produced in goat, affinity isolated antibody (Sigma-Aldrich), diluted 1:10,000 in 3% BSA solution.

6. BCIP/NBT substrate solution (Merck).

7. Silver staining kit or appropriate chemicals.

2.6 MS-Based Proteomics

UCSF protocol, slightly modified (http://msf.ucsf.edu/protocols.html).

All reagents and solution are prepared from MS grade components:

1. Destaining solution: 40% methanol, 50% water, 10% acetic acid.

2. IAM: iodoacetamide.

3. Trypsin (from porcine pancreas, proteomics grade, Sigma-Aldrich).

4. HCOOH: formic acid.

5. ACN: acetonitrile.

6. NH_4HCO_3: ammonium bicarbonate.

2.7 Cell Lysis

1. CHAPS: 3-[(3-Cholamidopropyl)dimethylammonio]-1-propanesulfonate.

2. HEPES: 4-(2-Hydroxyethyl)piperazine-1-ethanesulfonic acid.

3. EGTA: Ethylene glycol-bis(2-aminoethylether)-N,N,N', N'-tetraacetic acid.

4. EDTA: Ethylenediaminetetraacetic acid.

5. PIC: Protease inhibitor cocktail.

2.8 Instrumentation

1. UV-vis spectrophotometer, plate reader.

2. Shaking table.

3. Desalting spin columns 7K and 40K MWCO.

4. Amicon ultraconcentrator, 10K and 30K MWCO.

5. EDTA blood collection tubes with lavender conventional stopper.

6. Magnetic separator (Dynal MPC or analogue).

7. Mikro-Dismembrator.

8. C18 ZipTip pipette tips.

9. LC-MS instrument (*see* details in Subheading 3.4.4).

3   Methods

Perform all steps at room temperature unless otherwise specified.

3.1   Preparation of Persulfidating Polysulfide Reagent

Throughout the vast majority of our experiments, custom-made inorganic polysulfide reagent solution, a rich source of sulfane sulfur, was used to trigger persulfidation on proteins in biological samples [14]. Polysulfide reagent solutions were prepared by the previously characterized rapid oxidation of hydrogen sulfide with sodium hypochlorite [15].

3.1.1   Preparation and Handling of Hydrogen Sulfide Stock Solution

Prepare each stock solution freshly before use and store them on ice, protected from light. For safety purposes, use chemical fume hood when working with hydrogen sulfide stock solutions and confirm appropriate ventilation of the room. Discard leftover sulfide and polysulfide solutions into sodium hypochlorite solution.

1. Prepare DTNB stock solution: 4 mg/ml (10 mM) in 100 mM phosphate buffer, pH 7.4.

2. Take a larger (6–8 mm) crystal of $Na_2S \cdot 9H_2O$, and wash it several times with ultrapure water to remove polysulfide contamination from its surface (see **Note 1**).

3. Fully dissolve the crystal in 5–6 ml degassed ultrapure water (see **Note 2**). Store the stock solution on ice, and use it within 1–2 h $\rightarrow HS^-_{cc}$.

4. Make an appropriate dilution of HS^-_{cc} in ultrapure water (HS^-_{dil}), and measure its absorbance at 230 nm in a quartz cuvette. The pH of the HS^-_{dil} solution has to be at least 9–10, because for spectrophotometric detection purposes, sulfide needs to be in its single protonated HS^- form. To ensure this pH value without adding extra base or buffer, the concentration of the HS^-_{dil} solution should be 50–60 μM [16]. Under these conditions, the measured absorbance is between 0.4 and 0.5, because the molar absorption coefficient of HS^- at 230 nm is 7,700 M^{-1} cm^{-1}. Calculate the concentration of the HS^-_{dil} solution—$c_{HS^-,1}$:

$$c_{HS^-,1} = \frac{Abs_{230}}{7,700}$$

5. Add 1/10 volume unit of 10 mM DTNB to the sulfide solution in the same cuvette. An intense yellow product (TNB^-) is formed in 1–2 min; measure its absorbance at 412 nm, using $\varepsilon = 14,100$ M^{-1} cm^{-1}. Considering (1) the fact that two equivalents of TNB^- are produced in the reaction under these conditions [17] and (2) the dilution of the solution by the added DTNB, the applied formula is:

$$c_{HS^-,2} = \frac{Abs_{412}}{14,100 \times 2} \times 1.1$$

6. The applied concentration of the sulfide stock solution is calculated as the average obtained from the two methods (**steps 5** and **6**). If $c_{HS^-,1}$ and $c_{HS^-,2}$ differ by more than a few percent, the whole procedure should be repeated using fresh reagents (*see* **Note 3**).

$$c(HS^-_{cc}) = \frac{c_{HS^-,1} + c_{HS^-,2}}{2} \times \text{dilution factor}$$

3.1.2 Preparation of Polysulfide Stock Solution

Polysulfide reagents (either homemade or commercially available (*see* below)) are mixtures of polysulfur chains of different lengths [5]. For the sake of simplicity, the sum of inorganic polysulfide species (not taking into consideration their condition-mediated speciation) will be denoted as HS_x^-.

1. Prepare a 1–2 mM sodium hypochlorite solution in ultrapure water (*see* **Note 4**). Make sure that the pH of the solution is >9. Measure the absorbance of the solution at 292 nm, in a 1 cm path length cell using $\varepsilon_{292\ nm} = 350\ M^{-1}\ cm^{-1}$. Calculate the concentration of the stock solution using the formula below:

$$c(OCl^-_{cc}) = \frac{Abs_{292}}{350} \times \text{dilution factor}$$

2. Mix sulfide and hypochlorite stock solutions in ultrapure water at 10 mM OCl^- and 30 mM HS^- final concentrations. In order to avoid sulfur precipitation and overoxidation of HS^- to different oxysulfur species, add the OCl^- into the sulfide solution (not the other way around) dropwise during vigorous vortexing (*see* **Note 5**). Equal volume of the stock solutions should be used. The solution turns gradually yellow because of the formation of polysulfide species [15]. Keep the reagent on ice and protect from light. Use the polysulfide stock solution within 30 min.

3.1.3 Stability of Polysulfide Preparations

1. In Subheadings 3.1.1 and 3.1.2, the HS^- and HS_x^- stock solutions were prepared in ultrapure water. Because a Na_2S salt was used as a source of sulfide, these stock solutions have highly alkaline pH due to the dissolution process ($S^{2-} + H_2O \rightleftharpoons HS^- + OH^-$). Working dilutions of the polysulfide reagent can be made in the appropriate buffer, although the stability of the reagent at the commonly used pH 7.4 is a critical issue (Fig. 2a). First of all, the dilution buffers should at all times contain a chelating agent, such as DTPA or EDTA (DTPA is recommended), because trace amounts of transition metals in the buffers effectively catalyze the autoxidation of sulfide and polysulfide species. Furthermore, at lower pH

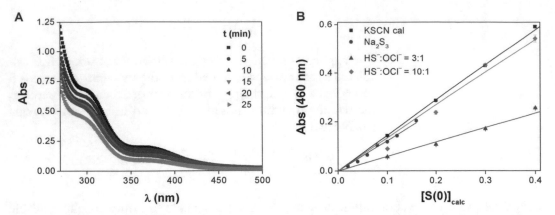

Fig. 2 Stability and sulfane sulfur content of different polysulfide preparations. (**a**) 20 mM polysulfide reagent was prepared by mixing OCl^- and HS^- stock solutions at 20 mM and 100 mM final concentrations, respectively. The strongly alkaline stock solution was diluted to 1 mM in HBSS (pH 7.40), and the diluted solution was stored at room temperature, protected from light. The spectrum of this solution was recorded at the indicated time points, taking a fresh aliquot each time to exclude photochemical interference. The intensity of the peak at 300 nm decreased by 36% over the indicated time scale. (**b**) Cold cyanolysis assay was carried out to compare the sulfane sulfur content of commercially available Na_2S_3 solutions (red circles, ●) and polysulfide reagent prepared from HS^- and OCl^- stock solutions, in 3:1 and 10:1 concentration ratios (blue triangles (▲) and green diamonds (♦), respectively). Black squares (■) represent the reference absorbance values obtained by using SCN^- standard solutions. The KSCN standards (■), Na_2S_3 samples (●), and the samples at 3:1 HS^- to OCl^- ratio (▲) were prepared in 200 mM Tris–HCl buffer, containing 100 μM DTPA at pH 7.40, followed by setting the pH to ~9–10 for the cold cyanolysis assay. The samples at 10:1 HS^- to OCl^- ratio (♦) were diluted in the same buffer at pH 9.60. The concentration of weighed Na_2S_3 was doubled to calculate the sulfane sulfur concentration of the corresponding samples

values, polysulfides undergo disproportionation reactions resulting in insoluble sulfur formation [15]. These disproportionation reactions are kinetically second-order processes for polysulfide concentrations, and therefore, it is not recommended to prepare polysulfide stock solutions at high concentrations (should preferably be <2 mM) due to more rapid sulfur precipitation. Due to the above reasons, it is also essential that the working dilutions are prepared right before use. As shown in Fig. 2b, the sulfane sulfur content of homemade polysulfide solutions were found adequate, but it is not recommended to go below a 4:1 = sulfide/OCl^- ratio due to overoxidation-induced formation of higher sulfur oxidation states containing species that are not capable of inducing Cys persulfidation.

2. Alternatively, commercially available polysulfide salts can be used to prepare the persulfidating solutions. We have tested sodium sulfide salt from Dojindo laboratories and confirmed their sulfane sulfur content using cold cyanolysis (Fig. 2b). Using this Na_2S_3 salt, it is recommended to stay with the concentrations of polysulfide stock solutions below 100 μM,

because our turbidimetric measurements indicated rapid sulfur precipitation above this concentration (data not shown).

3.2 Validation of the ProPerDP Method on Human Serum Albumin

Human serum albumin was chosen as a suitable model protein for validation of the ProPerDP method, because it only has one free, redox reactive cysteine residue. The other cysteine residues in the protein form 17 structural disulfide bonds. The persulfidation of HSA was previously demonstrated by an indirect method using monobromobimane labelling of reductively released hydrogen sulfide from the persulfidated sample, where it was confirmed that one cysteine equivalent is per-/polysulfidated upon polysulfide treatment of HSA [1, 14]. The polysulfide-induced persulfidation of HSA was shown to be fully reversible by reducing agents such as DTT or TCEP or in the presence of excess sulfide [1, 17].

1. Pre-reduce 8–10 mg/ml (120–150 μM) HSA with 1 mM TCEP for 30 min in TE buffer (*see* **Notes 6** and **7**).

2. Desalt with Thermo Zeba spin desalting column (7K or 40K MWCO).

3. Incubate pre-reduced sample with 1 mM polysulfide reagent for 30 min in the dark (Mix protein and 10 mM polysulfide stock solution in 9:1 volume ratio; this way the buffer concentration remains sufficient to set constant pH.)

4. Desalt with Thermo Zeba spin desalting column (7 K or 40 K MWCO).

5. Alkylate with 1 mM IAB for 1 h in the dark (*see* **Note 8**).

6. Desalt with spin column once and Amicon ultraconcentrator (30K) three times (*see* **Note 9**).

7. Determine the protein content of the desalted sample and dilute it to 0.33 mg/ml. Mix 27 μl of the desalted sample with 9 μl 4× SB and store at −20 °C. This sample represents *S1* in Fig. 1.

8. Wash 100 μl streptavidin-coated magnetic beads three times with equal volume of TE buffer, under short vortexing conditions each time.

9. Shake 27 μl alkylated sample with 100 μl previously washed streptavidin-coated magnetic beads (*see* **Note 10**) for 30 min using a shaking table (*see* **Notes 11** and **12**).

10. Separate the beads and the supernatant with a magnetic separator. The supernatant is *S2*, the non-alkylated fraction (Fig. 1). Put it in clean tubes, add sample buffer, and store at −20 °C.

11. Wash the beads three times with 1 ml of TTBS buffer, by short vortexing each time.

12. After the removal of the last washing buffer, resuspend the beads with 27 μl 5 mM TCEP, and shake for 30 min.

13. Separate beads from supernatant. Add 9 μl 4× SB to supernatant, this is the sample of interest, *S3*, which contains the persulfidated fraction (Fig. 1).

14. Boil the beads for 3 min at 100 °C in 36 μl 1× SDS-PAGE loading buffer. The supernatant is *S4*, representing the total thiol fraction (Fig. 1).

15. Discard magnetic beads (*see* **Note 13**).

16. Run 20 μl of *S3* (from **step 13**) samples on 12% SDS-PAGE gel. This volume represents the persulfide fraction of 5 μg total protein in *S1* (*see* **Note 14**).

17. Stain gel with colloidal Coomassie dye.

Representative gels showing HSA persulfidation are shown in Fig. 3.

3.3 Application of the ProPerDP Method on Complex Biological Samples/Materials

3.3.1 Polysulfide Dose Dependence of HSA Persulfidation in Human Blood Plasma

The persulfidation of serum albumin can be shown in biological context too, such as in human blood plasma, where HSA was previously suggested as a H_2S carrier [18, 19].

1. Collect peripheral venous blood from healthy adult human volunteers with informed consent into EDTA collection tubes (*see* **Note 15**).

2. Centrifuge the blood sample at 3000 × *g* for 10 min.

3. Aliquot the supernatant plasma into equal volumes and measure protein content (*see* **Note 16**).

4. Dilute samples to 10 mg/ml total protein content.

5. Treat the samples with 0–1 mM polysulfide reagent for 30 min in the dark (*see* **Note 17**).

6. Repeat **steps 4–15** from Subheading 3.2. In **step 7**, dilute the samples to 10 ng/μl in TE buffer.

7. Run *S3* samples on 12% SDS-PAGE gel; loading *S3* samples correspond to the persulfide pool of 90 ng total proteins in the *S1* samples/lane (12 μl).

8. Perform Western blotting against HSA, according to Subheading 3.4.2. *See* representative Western blot results in Fig. 3c, c′.

3.3.2 Persulfidated Proteins in Polysulfide-Treated Cell Lysate

The example is provided for A549 cells based on the reported data in Fig. 4a from [1].

1. Culture A549 cells (ECACC) in DMEM-F12 medium (Lonza) supplemented with 10% (v/v) heat-inactivated fetal bovine serum (FBS), penicillin (100 units/ml), and streptomycin (100 μg/ml) (Sigma) in 175 cm² culture flask until 80–90% confluency (*see* **Notes 18** and **19**).

2. Seed the cells with trypsin-EDTA solution and count the total cell number.

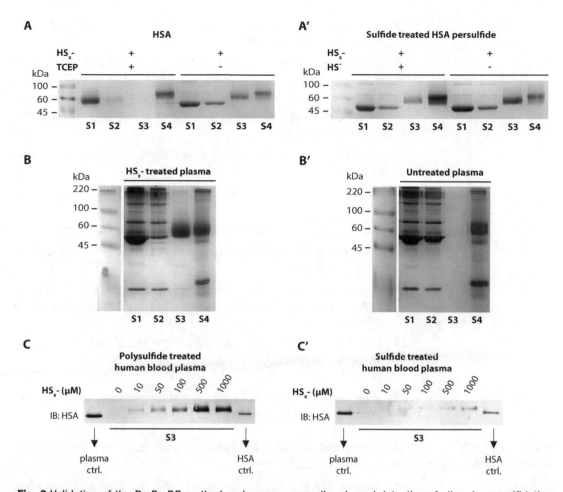

Fig. 3 Validation of the ProPerDP method on human serum albumin and detection of albumin persulfidation from human blood plasma (reproduced from [1] with permission from AAAS). (**a**) and (**a'**) Pre-reduced HSA samples were treated with polysulfide reagent to induce per-/polysulfide formation on the Cys34 residue. The samples were alkylated by IAB and pulled down using streptavidin-coated magnetic beads. *S1* and *S2* lanes confirm the efficiency of the pulldown step. The *S3* lanes represent the per-/polysulfidated fractions and *S4* the total thiol pool of the samples. (The *S1–S4* labelling of the samples follows the order of Fig. 1 on each panel.) (**a**) TCEP and (**a'**) HS⁻ treatment corroborated the reversible nature of persulfide formation. 5 μg total protein was loaded in the *S1* lanes, and gels were stained by colloidal Coomassie staining. MS identification confirmed that all the lanes contain human serum albumin. (**b**) and (**b'**) Polysulfide treatment of human blood plasma samples leads to significant extent of per-/polysulfide formation. We were unable to detect protein polysulfidation in the untreated plasma samples under the applied conditions. 15 μg total protein was loaded in the *S1* lanes, and gels were stained by colloidal Coomassie staining. (**c**) and (**c'**) Albumin per-/polysulfidation was detected from diluted plasma samples by Western blot analysis. Only the *S3* samples were loaded on the gels (the persulfide fraction of 90 ng total serum protein), along with 100 ng plasma protein and 20 ng purified HSA controls. The mobility shift on the gels and blots is the consequence of the intramolecular disulfide bond reduction in the protein structure after protein denaturation. (**c**) The extent of polysulfide-induced albumin per-/polysulfidation increased in a polysulfide reagent concentration-dependent manner. (**c'**) Weak signals were observed in the corresponding control experiments of (**c**) where only hydrogen sulfide treatment was applied, which was assigned to the actions of polysulfide contamination in sulfide stock solutions

3. Centrifuge the cell suspension at $14,500 \times g$ for 5 min and discard the supernatant.

4. Incubate the cells in 600 μl cell lysis buffer for 5–10 min at 4 °C.

5. Centrifuge the lysate at $14,500 \times g$ for 5 min at 4 °C, and transfer the supernatant into a clean tube, and discard the insoluble fraction.

6. Determine the total protein concentration of the lysate (*see* **Note 20**).

7. Treat the lysate with 3 mM polysulfide for 30 min (*see* **Note 21**).

8. Desalt the sample with desalting spin column (*see* **Note 22**).

9. Alkylate the flow-through with 5 mM IAB for 3 h in the dark (*see* **Note 23**).

10. Repeat **steps 6–15** from Subheading 3.2. In **step 7**, dilute samples to 1 mg/ml, and incubate 21 μg protein with 250 μl of beads.

11. Run 20 μl of *S3* samples on 12% SDS-PAGE gel. This represents the persulfide fraction of 10 μg total protein in *S1* (*see* **Note 14**).

12. Stain gel with colloidal Coomassie stain (Subheading 3.4.1) or perform silver staining according to Subheading 3.4.3.

3.3.3 Exploring the Persulfidome in Intact Cells

The example is provided for HEK293 cells based on the reported data in Fig. 5 from [1].

1. Culture HEK293 cells (ATCC) in Eagle's minimum essential medium, complemented with 10% (v/v) FBS, penicillin, (100 U/ml), and streptomycin (100 mg/ml). Cells were kept in logarithmic growth phase at 37 °C in humidified air containing 5% CO_2 in 175 cm^2 culture flask until 80–90% confluency (*see* **Notes 18** and **19**).

2. Discard medium.

3. Wash the cells with HBSS.

4. Expose the cells to 200 μM polysulfide in HBSS for 2 h (*see* **Note 24**).

5. Wash the cells with HBSS (*see* **Note 24**).

6. Alkylate the cells with 1 mM IAB diluted in HBSS for 3 h (*see* **Notes 25–27**).

7. Scrape the cells into the alkylating solution, and centrifuge the suspension at $14,500 \times g$ for 5 min at 4 °C.

8. Discard supernatant.

9. Add 500 µl cell lysis buffer to the cell pellet, suspend, and vortex. Incubate for 5–10 min.

10. Centrifuge the lysate at 14,500 × g for 5 min at 4 °C.

11. Transfer supernatant into clean tube.

12. Measure protein concentration and set protein level to 1 mg/ml.

13. From this point, perform ProPerDP assay according to **steps 8–15** in Subheading 3.2, considering Subheading 3.4.3 for silver staining and Subheading 3.4.4 for MS proteomics analysis, or Western blot analyses for identification of a single protein, depending on the purpose of the experiment.

3.3.4 Detection of Persulfidation in Frozen Tissue Samples

1. Perfuse mouse livers (or organ of interest) by a cardiac-to-portal route with a solution of 150 mM NaCl, 100 mM pipes (pH 7.1), and 3.0 mM IAB (*see* **Note 28**).

2. Remove perfused livers and divide into ~100 mg pieces.

3. Submerge each piece in a 1.5 ml tube containing 65 µl perfusion buffer.

4. Incubate the samples on ice for 15–20 min.

5. Remove excess perfusion buffer, seal the tubes, and snap-freeze the samples in liquid nitrogen.

6. Keep the samples at −80 °C or on dry ice until use.

7. Drop frozen tissue samples into liquid nitrogen immediately after taking them out from −80 °C. Make a fine powder from these samples in a liquid nitrogen-precooled Teflon/iridium carbide ball Mikro-Dismembrator 2 ball mill (B. Braun Melsungen AG) (*see* **Note 29**).

8. Add 3 mM IAB solution to this fine powder, transfer the mixture into clean centrifuge tube, and incubate the mixture for 1 h in the dark at RT.

9. Add 1% CHAPS and 1% protease inhibitor cocktail (*see* **Note 30**), and incubate the mixture for an additional 30 min.

10. Centrifuge the suspension at 14,500 × g for 10 min at 4 °C.

11. Remove debris and desalt the supernatant carefully by desalting spin column (once) and Amicon ultraconcentrator (three times) (*see* **Note 22**).

12. Measure the protein content of the supernatant by the Bradford method, and set the protein levels to 1 mg/ml.

13. From this point, perform ProPerDP assay according to **steps 8–15** in Subheading 3.2, considering Subheading 3.4.3 for silver staining and Subheading 3.4.4 for MS proteomics analysis, depending on the purpose of the experiment.

3.4 Detection Methods for the Separated Persulfide Pool

The persulfidated proteins isolated by the ProPerDP method (*S3* in Fig. 1) can be analyzed by SDS-PAGE and downstream Coomassie staining, Western blotting, silver staining, and LC-MS proteomics methods, depending on the purpose of the experiment. Although these procedures are commonly known in the field, a short note is given below with regard to their joint application with the ProPerDP method, and detailed protocol is provided for MS proteomics analysis. It's noteworthy that the sensitivity of the chosen detection method determines the total amount of protein to be used in the experiments and consequently the volume of streptavidin beads required per sample. The indicated protein-to-bead ratios are based on estimations from the beads' binding capacities and on our own trial-and-error-based method optimizing experience for the corresponding samples.

3.4.1 Coomassie Staining

Any commercially available Coomassie dye is applicable with no special reference. However, colloidal Coomassie dye has higher sensitivity, and the destaining step prior to MS proteomics analysis is facilitated compared to the regular Coomassie Brilliant Blue dyes. The total protein contents of the corresponding S1 samples were 5 µg for HSA (Fig. 3a), 15 µg for plasma (Fig. 3b, b'), and 10 µg for A549 cell lysate (*see* Fig. 4a in [1]).

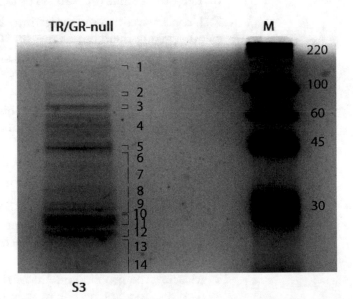

Fig. 4 Detection of Cys sulfhydration from TR/GR-null mouse liver. The figure shows a representative gel on the protein persulfide pool of a TR/GR-null mouse liver sample. The persulfide fraction of 250 µg total protein is loaded on the gel and stained by colloidal Coomassie staining. The indicated bands were isolated for MS protein identification, which resulted in 280 peptide identification from 98 unique proteins

A special case is the MS proteomics identification of the persulfidome, where our instrumentation required Coomassie staining and higher amount of total protein; *see* detailed discussion below in Subheading 3.4.4.

3.4.2 Western Blot Against HSA

The procedure below refers to the commonly used conditions in our lab. It can be substituted by any appropriate immunoblotting protocol against human serum albumin.

1. After gel electrophoresis, transfer the samples to PVDF membranes.

2. Block overnight in 3% BSA solution at 4 °C.

3. Incubate membrane with primary antibody solution for 1 h at room temperature with gentle shaking.

4. Wash membrane for 30 min in TTBS buffer.

5. Incubate membrane with secondary antibody solution for 1 h at RT.

6. Wash membrane for 30 min in TTBS buffer.

7. Add 2.5 ml of BCIP/NBT ready-to-use alkaline phosphatase substrate solution (*see* **Notes 31** and **32**).

8. Wash membrane with ultrapure water when the required signal intensity is reached.

3.4.3 Silver Staining

Due to its higher sensitivity compared to Coomassie staining, silver staining was proved effective in detecting endogenous persulfidation levels in intact cells (A549 and HEK293) and tissue samples. Within its linearity range, silver staining is suitable to semiquantitatively compare persulfidation in parallel samples, such as different cell lines, treatments, or tissues with diverse genetic background. Because of the autocatalytic nature of the reduction of silver [20], samples have to be run on the same gel to be able to make a comparison (*see* **Note 33**). By loading appropriate dilutions of the corresponding *S1* sample (representing the quantified total protein pool of the sample) on the gel (e.g., 100-, 200-fold), approximation can be made on the concentrations of their persulfide fractions (*S3*) based on relative densitometry.

Any commercially available silver staining kit or a set of homemade silver staining solutions are appropriate for this step. We adapted an MS compatible, highly sensitive protocol for two dimensional gels. The persulfide pool of 20–30 µg total protein in an aliquot of cell or tissue lysate is normally readily visible by silver staining (*see* **Note 34**). An example for the conditions optimal for silver staining for the analogue steps of **steps 7–16** in Subheading 3.2 is given below:

1. Dilute cell or tissue homogenate samples to 1 mg/ml total protein content.

2. Wash 500 μl streptavidin-coated magnetic beads with equal volume of TE buffer three times, by short vortexing each time.

3. Mix 60 μl of cell homogenate with 20 μl of 4× SB, and store at −20 °C → *S1* (Fig. 1).

4. Mix 60 μl of cell homogenate (60 μg total protein) with 500 μl beads, and shake for 30 min on a shaking table.

5. Separate the beads from the supernatant with a magnetic separator, and transfer the supernatant into clean tube. Add 4× SB → *S2* (Fig. 1).

6. Wash the beads three times with 1 ml TTBS by short vortexing each time.

7. Resuspend the beads in 60 μl 5 mM TCEP and shake for 30 min on a shaking table. After magnetic separation, add 4× SB to the sample → *S3* (Fig. 1) (*see* **Note 35**).

8. Boil the beads in 1× SB for 3 min at 100 °C → *S4* (Fig. 1).

9. Run 35 μl of *S3* samples on 12% SDS-PAGE gel. This represents the persulfide fraction of 26 μg total protein in *S1* (*see* **Note 14**).

10. Perform silver staining (*see* **Note 36**).

3.4.4 MS Proteomics Analysis

An example for the conditions optimal for LC-MS analysis for the analogue steps of **steps 7–16** in Subheading 3.2 is given below:

1. Incubate 250 μg total protein with 2 ml Dynabeads® M280 (Life Technologies) in four equal portions (*see* **Notes 37** and **38**) for 30 min on a shaking table.

2. Transfer supernatant into clean tubes → *S2*.

3. Wash beads six times with 1 ml TE buffer (*see* **Note 39**).

4. Before the removal of the last washing buffer, pool the four samples together, separate the beads with a magnet, and discard all the supernatant.

5. Resuspend the beads in 45 μl 5 mM TCEP, and shake for 30 min on a shaking table (*see* **Note 40**).

6. Separate the beads with a magnet, and transfer supernatant to a clean tube, and add 4× sample buffer → *S3*.

7. Run 55 μl of the sample on a 1.5 mm SDS-PAGE gel.

8. Stain the gel with colloidal Coomassie stain.

Use MS grade solvents and chemicals, and perform extreme care for purity throughout the proteomics analysis, in order to minimize contamination of the samples, for example, by ubiquitous skin keratin from dust or personnel.

Perform MS proteomics analysis according to UCSF in-gel digestion protocol with some modifications [21, 22], according to the procedure below.

Day 1: Dissection and destaining

1. Dissect bands of interest or the entire lane into equal pieces and cut into 1×1 mm cubes.

2. Destain the dissected bands in destaining solution: 40% MeOH, 10% acetic acid, and 50% water. Incubate the gel cubes in the destaining solution overnight, and if they still look pale blue, add fresh destaining solution for another few hours.

Day 2: Tryptic digest

3. Add 100 μl of 25 mM NH_4HCO_3 in 50% ACN solution to the gel pieces and vortex for 10 min. Repeat two times (*see* **Note 41**).

4. Reduce the gel cubes in 100 μl of 10 mM DTT in 25 mM NH_4HCO_3 at 56 °C for 1 h (*see* **Note 42**).

5. Remove supernatant, and then alkylate the gel cubes in 100 μl of 55 mM iodoacetamide in 25 mM NH_4HCO_3 for 45 min at RT in the dark.

6. Remove supernatant and wash gels with 100 μl 25 mM NH_4HCO_3 for 10 min.

7. Dehydrate gel pieces with 100 μl of 25 mM NH_4HCO_3 in 50% ACN two times.

8. Dry gel pieces with 100 μl ACN and discard supernatant.

9. Add trypsin solution to just barely cover the gel pieces (5–25 μl) (*see* **Note 43**).

10. Rehydrate the gel pieces on ice or at 4 °C for 10 min. Spin briefly. Add 25 mM NH_4HCO_3 as needed to cover the gel pieces (30 μl).

11. Spin briefly and incubate overnight (minimum 16 h) at 37 °C.

Day 3: Extraction of peptides

12. Transfer the digest solution (aqueous extraction) into a clean 1.5 ml siliconized tube.

13. To the gel pieces, add 30 μl (enough to cover) of 50% ACN/5% formic acid, vortex for 20 min, spin down, and sonicate for 10 min. Repeat once.

14. Vortex the extracted peptides and spin down.

15. Prepare the following solutions for isolating digested peptides with C18 ZipTips (Millipore):

    (a) Wetting solvent: 100% ACN (LC-MS grade).

      (b) Equilibration/wash solution: 0.1% HCOOH/H$_2$O (LC-MS grade).

      (c) Elution solution: 0.1% HCOOH/50% ACN–50% H$_2$O.

16. Pre-wet C18 ZipTips (Millipore) by aspirating and dispensing 10 μl ACN, and discard ACN to waste. Repeat 3–4 times.

17. Equilibrate C18 ZipTips by aspirating and dispensing 10 μl equilibration solution. Repeat 3–4 times.

18. Bind-digested peptides on C18 ZipTips by aspirating and dispensing sample on the tips 10–15 times (*see* **Note 44**).

19. Wash C18 ZipTips by aspirating and dispensing 10 μl wash solution. Repeat 3–4 times.

20. Pipette 20 μl of elution solution into clean Eppendorf or 0.2 ml PCR tubes for each sample.

21. Elute the peptides into the previously aliquoted elution solution by aspirating and dispensing it 10–15 times, keeping the tip in the solution. If several ZipTips were used per sample, elute each tip in the same tube. Discard used ZipTips.

22. Inject 10 μl of cleaned peptide solution to LC-MS instrument or store at −80 °C until injection.

23. The technical parameters of our LC-MS system are described below (they can be appropriately adapted to the instrument locally available).

    Chromatographic separation of peptides:

      (a) Reverse phase: nanoflow HPLC (Easy-nLC II, Thermo Fisher Scientific).

      (b) Column: EASY-column (10 cm, 75 μm, C18) 5%.

      (c) Eluent: gradient elution (40 min/run).

      (d) Start: 5% (v/v) ACN, 0.1% (v/v) HCOOH in water.

      (e) End: 90% (v/v) ACN, 0.1% (v/v) HCOOH in water over.

      (f) Flow rate: 300 nl/min.

    Detection:

    LTQ XL ion trap mass spectrometer (Thermo Fisher Scientific).

    Settings:

      (a) Triple play positive ion mode.

      (b) Mass range (m/z): 400–1500.

    Peptide Identification:

      (a) Protein Prospector (UCSF Mass Spectrometry Facility, v 5.14.1).

(b) Database: NCBI nr.2013.06.17.

(c) Taxonomy: sample related.

A representative Coomassie-stained gel from TR/GR-null mouse liver [23], which lack both thioredoxin reductase-1 and glutathione reductase, is shown in Fig. 4.

3.5 Limitations of the ProPerDP Method

The methodology provided herein is not without caveats, which should be taken into consideration using it for protein persulfide detection.

In the case of proteins having multiple surface-exposed Cys residues, it is conceivable that the protein has a persulfidated and a reduced cysteine as well. In this case both residues get labelled in the alkylating step and attached to the streptavidin beads in the subsequent pulldown. However, even if the resulting disulfide bond (from the persulfidated Cys) gets reduced, the protein remains on the beads, because it is harbored by the thioether bond (coming from the alkylation of the reduced Cys residue) (Fig. 5a). Furthermore, false-positive signals can arise from

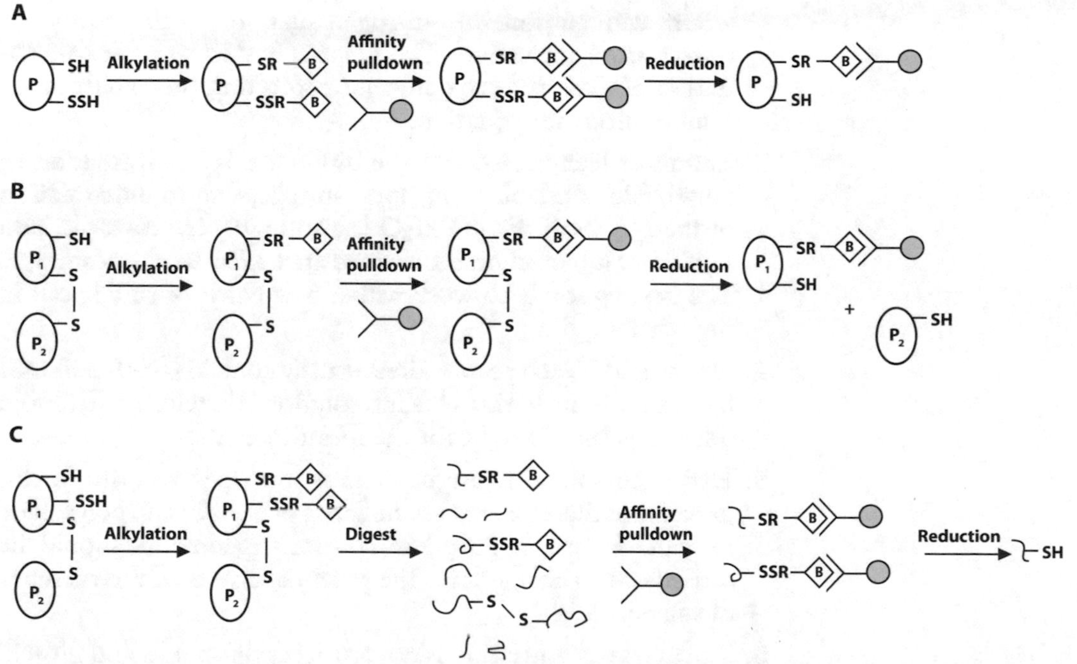

Fig. 5 Caveats of the ProPerDP method and a suggested solution (reproduced from [1] with permission from AAAS). (**a**) Multiple cysteine proteins might exert false-negative signal in the persulfide sample (*S3*) if more than one surface-exposed cysteine residue is available for alkylation, but not all of them is prone to per-/polysulfide formation. Such a structure leads to a reduction-resistant attachment of the protein to the beads as shown in the figure. (**b**) Intermolecular disulfide bond formation in a protein mixture can cause a false contribution to the persulfide signal due to the cleavage of the disulfide bond after the pulldown step. (**c**) Performing the biotin-streptavidin pulldown on the peptide level following tryptic digest of the protein mixture and subsequent LC-MS analysis is a potential solution to the drawbacks outlined in (**a**) and (**b**). This addition to the methodology and its optimization is currently underway in our laboratory

intermolecular disulfide bonds existing in the protein mixture. Figure 5b shows a case where a protein without any persulfidated cysteine appears in the *S3* sample because of the reduction of an intermolecular disulfide bond during the reduction step. However, our previous diagonal gel-based experiments suggested that the contribution of this factor is minor compared to the overall persulfide signal [1]. Figure 5c indicates that carrying out the pulldown step after tryptic digest could overcome the abovementioned situations. A protocol for this has since appeared in the literature [24].

4   Notes

1. 120 mg/5 ml makes a 100 mM stock solution by weight, assuming 100% purity, although the concentration has to be determined because of potential contaminations. The quality of the sulfide donor is a crucial point. A description of different sulfide donors and handling of sulfide reagents is given in a previous report from our laboratory [16].

2. Several degassing options are applicable, like bubbling the water with argon or nitrogen gas of high purity or low-pressure sonication. The point is to decrease the concentration of dissolved oxygen, thus protecting the sulfide stock solution from autoxidation.

3. The most likely reason of the difference is the formation of polysulfide contamination upon storage, due to autoxidation of the Na_2S salt. $Na_2S \cdot 9H_2O$ is also highly hygroscopic, and significant liquefaction is observed after a few weeks of storage. The best option is to work with a new batch of sulfide salt in such cases.

4. Household bleach preparations usually contain ~100–200 mM hypochlorite in highly alkaline solution; therefore a 100-fold dilution is usually right for the measurement.

5. Hydrogen sulfide is used in excess over hypochlorite in this procedure; therefore excess sulfide is present in the polysulfide reagent. Hence, appropriate control experiments should be carried out to characterize the potential effects of the presence of sulfide.

6. The exact concentration is not crucial at this point, and protein concentration should be precisely set before the pulldown step.

7. The purpose of the prereduction step is to reduce the redox active Cys34 into the active thiol form, from partially oxidized forms such as CysSOH or CysS-SG [25]. Experimental data suggest that the prereduction doesn't affect the disulfide bonds in the protein structure. The mere fact that serum albumin persulfide is detected on the Coomassie-stained gels shows

that no further reactive cysteines would be produced upon prereduction (Fig. 5a). This is in line with the monobromobimane assay showing approximately one HSA equivalent of sulfide is liberated from a polysulfide treated HSA sample.

8. Iodoacetyl-PEG2-biotin is readily soluble in aqueous buffers, although it is sensitive to moisture and light. Let the vial—containing the solid reagent—warm to room temperature for 15–20 min before use, and protect it from light, as well as the prepared solutions. Prepare IAB solutions right before use, and check its pH with a small piece of indicator paper. It tends to shift pH toward the acidic range.

9. Extensive desalting is essential to maximalize the efficiency of the upcoming streptavidin pulldown. Free IAB molecules, being smaller than biotinylated protein ligands, can easily saturate the binding sites of the streptavidin-coated beads. Assuming a 99% efficacy for the desalting steps would account for 10 μM remaining free IAB in the sample under the applied conditions, which is comparable to the protein concentration. Another possibility is to use an excess of beads to overcome the residual free IAB in the sample; however, this latter option is unfavorable from a financial point of view. The streptavidin beads usually represent the main cost item in the whole experiment.

   In order to maximize the desalting efficiency, we optimized the performance of the spin columns before use using the fluorescent signal of 5-iodoacetamido fluorescein (IAF) as indicator of desalting level as well as measuring protein content before and after the purification step with the column to check protein recovery.

10. The total volume of the streptavidin-coated magnetic beads to be used is determined by the amount of protein undergoing affinity pulldown (which is estimated by the detection limit of the chosen detection method (*see* Subheading 3.4)) and the binding capacity of the magnetic beads (which is provided by the manufacturer). According to the related product information, the binding capacity of Dynabeads® M280 is 650–900 pmol free biotin/mg beads (always check the product information sheet; binding capacity can differ between different lots of the same product). The concentration of the same product is 10 mg/ml (meaning 1% w/v solid content); therefore the binding capacity is 6.5–9 nmol free biotin/ml bead.

   For a general experiment when the persulfide pool was detected by silver staining, the pulldown step was carried out starting with 60 μg total protein. Assuming an average protein molecular weight of 40 kDa (in a cell or tissue homogenate) and 380 amino acid units/protein (ref: https://www.ocf.

berkeley.edu/~asiegel/posts/?p=7, based on NCBI protein library data), 60 μg protein refers to 3.5×10^{17} amino acid units. Based on published data, 2.26% of all the amino acid units is cysteine in mammals (the ratio is even smaller in organisms of lower level of complexity) [26], which means 7.91×10^{15} Cys units. This corresponds to 13.1 nmol free biotin, assuming that only thiol side chains are affected by IAB alkylation. The efficiency of the beads is around 50% (both experimental experience from protein measurement in *S1* and *S2* samples and from the *Technical Handbook* of Dynal, the former provider of Dynabeads® M280). Therefore, we are expected to pull down 6.6 nmol of biotinylated species. Regarding the abovementioned binding capacity, even in the best case scenario, this amount requires at least 1–1.5 ml bead suspension. We usually used 500 μl magnetic beads for 60 μg total protein, which is an underestimation based on the argument above, although meticulous control experiments suggested that with this protein-to-bead ratio, the experiments are properly reproducible, and the pulldown efficiency is equivalent between parallel samples of the same protein content.

Sigma-Aldrich provides similar value of biotin binding capacity for the related product used in our studies (Microparticles, magnetic, streptavidin coated; >600 pmol biotin-FITC/mg).

Depending on the purpose of the experiment (from Western blotting against specific protein to proteomics analysis of a complex mixture), this ratio can be used as a starting point to up- or downscale the used amounts of beads and total protein.

11. Alternatively, streptavidin agarose resin can be used for the pulldown step; however it needs to be noted that we performed all experiments with magnetic beads, except for one example, shown in Fig. S4 of [1].

12. Instead of a shaking device, a tube rotator can be used to ensure sufficient mixing of the beads with the protein sample. In fact, when the tubes are agitated horizontally, it should be checked in every few minutes that the beads are not settled in the bottom of the tube.

13. It has been reported that the biotin-streptavidin bond can be broken by simply heating the beads in water over 70 °C [27]. We have not verified this in our protocols and, instead, always use fresh beads for our pulldown experiments.

14. In practice, the total protein content of the *S3* sample is not precisely known, because it is affected by the efficiency of the persulfidation. Therefore, the total protein content of the *S1* sample is given, and the amounts in the *S3* lanes are referred to

as its persulfide fraction. 20 μl of the sample prepared in **step 7** contains 5 μg total protein (*S1*) by calculation.

15. Follow the relevant ethical regulations of experimentation involving human subjects.

16. Fresh plasma samples were used in our experiments. Frozen samples can be used replacing **steps 1–3**; however persulfidation status could be different in stored samples (not tested).

17. Prepare appropriate dilutions of the polysulfide reagent in ultrapure water right before mixing with protein sample, and protect from light. Add to the sample in 1:9 volumetric ratio, to minimize the drop in buffer concentration.

18. Our cells were checked for mycoplasma contamination by Mycosensor QPCR Assay Kit (Agilent).

19. Other cell lines can be used, but the conditions need to be optimized in this case.

20. Always check the chemical compatibility of the detergent with the applied protein detection assay.

21. **Steps 7** and **8** in Subheading 3.3.2 are optional. Omitting the polysulfide treatment and the subsequent desalting would provide information on the endogenous persulfide level in crude cell lysate. Use silver staining as detection method in this case.

22. Always check the compatibility of the desalting devices with the used detergents, both spin columns and Amicon ultraconcentrators.

23. These conditions were used during the optimization phase of the method; we recently changed them to 1 mM IAB for 1 h.

24. **Steps 4** and **5** in Subheading 3.3.3 are optional. Omitting the polysulfide treatment and the subsequent washing step would provide information on the endogenous persulfide level in intact cells. Use silver staining as detection method in this case.

25. The main advantage of our method is that our labelling agent is cell permeable; therefore the alkylation step can be performed on intact cells, before the lysis step. The persulfide level can be locked in an alkylated state inside the intact cell, eliminating/reducing concerns about artifact oxidative effects during the lysis step.

26. Besides the streptavidin-coated beads, IAB treatment of cells in culture flasks is another significant cost item of the experiments. Usually 20–25 ml of solution is used to cover the cells in a 175 cm^2 flask, but for economical reasons, 5 ml solution is sufficient for the alkylation step.

27. HEK293 cells are prone to detachment during different exposures, and we experienced significant detachment over the IAB labelling. The condition of the cells should be monitored

throughout the alkylation step, and if detachment is observed, then remove the buffer immediately, and add 500 μl lysis buffer to the flask directly, which contains 1 mM IAB. Lyse the cells as usually, and carry out careful desalting on the lysate (Spin column + Amicon ultraconcentrator three times).

28. All in vivo experiments must respect and carefully follow the relevant ethical guidelines for animal care.

29. The Mikro-Dismembrator 2 ball mill (B. Braun Melsungen AG) is no longer readily available on the market. Many laboratories use alternative techniques to prepare tissue homogenates (e.g., Physcotron tissue homogenizer and sonication). The advantage of the liquid nitrogen frozen ball mill method is that the tissue is powdered and alkylated while still in frozen state.

30. Use cell/tissue lysis buffer without PIC for MS proteomics experiments.

31. The primary antibody is highly sensitive; in optimal case, the signal appears almost immediately after adding the terminal developing solution.

32. When using the AP conjugated secondary antibody and BCIP/NBT for ultimate detection, the membranes tend to slightly darken while drying. Take image of the membrane and perform densitometry only when the blot is fully dried. This also minimizes the background signal from wet membrane.

33. If parallel samples are compared, make sure the total protein content is equal in the different *S1* samples.

34. Silver staining should be carried out with extreme care for purity of the solutions and cleanliness of the equipment.

35. This sample volume is sufficient for two runs on a standard 1.0 mm gel; the indicated quantities can be halved accordingly.

36. Stop the developing step of silver staining as early as possible, immediately as the signal in the loaded lanes starts to appear. This is a way to avoid the saturation of the signal and background on the gel.

37. For ESI-MS proteomics analysis, Dynabeads® M280 streptavidin-coated beads are recommended. In our experience, beads from another popular supplier gave a polymer signal in the mass spectrum, which suppressed the peptide signals and made the protein identification unreliable.

38. The total protein amount was increased to meet the sensitivity of our LC-MS instrumentation. The amounts indicated in **step** 7 in Subheading 3.2 were quadrupled, and 250 μg protein was incubated with 2 ml of Dynabeads® M280.

39. The usage of Tween 20 in the washing buffer should be avoided. Therefore the beads were washed six times in TE buffer before the reduction step instead of three times in TTBS.

40. The reduction step is an opportunity to concentrate our samples. The minimum volume of reducing buffer is a sufficient amount to cover the beads. The solid content of 2 ml Dynabeads® M280 can be covered in 45 μl of liquid in the bottom of a 1.5 ml Eppendorf tube.

41. Use gel loading tips to aspirate solutions from above the gel pieces.

42. At the beginning of each treatment, vortex the samples and spin briefly.

43. Trypsin solution is prepared according to manufacturer's recommendation (Sigma-Aldrich, trypsin from porcine pancreas, proteomics grade). Prepare a solution by adding 100 ml of 1 mM HCl to one 20 mg vial of trypsin. Mix the vial briefly to ensure the trypsin is dissolved. Add 900 ml of a 40 mM ammonium bicarbonate in 9% acetonitrile solution to the vial and mix. The final concentration of trypsin is 20 mg/ml.

44. 10 μl multichannel pipettes can be used for wetting and equilibrating, but single channel pipette is more accurate for sample binding and elution steps. Use one ZipTip for each 10 μl sample.

Acknowledgments

Financial support from the Hungarian National Science Foundation (OTKA; grant no.: K109843, KH17_126766, and K18_129286) for P.N.; from the National Institutes of Health (grant no.: R21AG055022-01) for E.E.S., P.N., and E.S.J.A.; and from the Swedish Research Council, Swedish Cancer Society, and Karolinska Institutet for E.S.J.A. is acknowledged. P.N. is a János Bolyai Research Scholar of the Hungarian Academy of Sciences. Dojindo Molecular Technologies Inc. is greatly acknowledged for their kind support of chemical supplies.

References

1. Doka E, Pader I, Biro A, Johansson K, Cheng Q, Ballago K, Prigge JR, Pastor-Flores D, Dick TP, Schmidt EE, Arner ES, Nagy P (2016) A novel persulfide detection method reveals protein persulfide- and polysulfide-reducing functions of thioredoxin and glutathione systems. Sci Adv 2(1):e1500968. https://doi.org/10.1126/sciadv.1500968

2. Mustafa AK, Gadalla MM, Sen N, Kim S, Mu WT, Gazi SK, Barrow RK, Yang GD, Wang R, Snyder SH (2009) H$_2$S signals through protein S-sulfhydration. Sci Signal 2(96):ARTN ra72. https://doi.org/10.1126/scisignal.2000464

3. Nagy P (2015) Mechanistic chemical perspective of hydrogen sulfide signaling. Methods

Enzymol 554:3–29. https://doi.org/10.1016/bs.mie.2014.11.036

4. Ida T, Sawa T, Ihara H, Tsuchiya Y, Watanabe Y, Kumagai Y, Suematsu M, Motohashi H, Fujii S, Matsunaga T, Yamamoto M, Ono K, Devarie-Baez NO, Xian M, Fukuto JM, Akaike T (2014) Reactive cysteine persulfides and S-polythiolation regulate oxidative stress and redox signaling. Proc Natl Acad Sci USA 111(21):7606–7611. https://doi.org/10.1073/pnas.1321232111

5. Cuevasanta E, Moller MN, Alvarez B (2016) Biological chemistry of hydrogen sulfide and persulfides. Arch Biochem Biophys. https://doi.org/10.1016/j.abb.2016.09.018

6. Yadav PK, Martinov M, Vitvitsky V, Seravalli J, Wedmann R, Filipovic MR, Banerjee R (2016) Biosynthesis and reactivity of cysteine persulfides in signaling. J Am Chem Soc 138 (1):289–299. https://doi.org/10.1021/jacs.5b10494

7. Millikin R, Bianco CL, White C, Saund SS, Henriquez S, Sosa V, Akaike T, Kumagai Y, Soeda S, Toscano JP, Lin J, Fukuto JM (2016) The chemical biology of protein hydropersulfides: studies of a possible protective function of biological hydropersulfide generation. Free Radic Biol Med 97:136–147. https://doi.org/10.1016/j.freeradbiomed.2016.05.013

8. Bianco CL, Chavez TA, Sosa V, Saund SS, Nguyen QN, Tantillo DJ, Ichimura AS, Toscano JP, Fukuto JM (2016) The chemical biology of the persulfide (RSSH)/perthiyl (RSS.) redox couple and possible role in biological redox signaling. Free Radic Biol Med 101:20–31. https://doi.org/10.1016/j.freeradbiomed.2016.09.020

9. Bailey TS, Pluth MD (2015) Reactions of isolated persulfides provide insights into the interplay between H2S and persulfide reactivity. Free Radic Biol Med 89:662–667. https://doi.org/10.1016/j.freeradbiomed.2015.08.017

10. Ono K, Akaike T, Sawa T, Kumagai Y, Wink DA, Tantillo DJ, Hobbs AJ, Nagy P, Xian M, Lin J, Fukuto JM (2014) Redox chemistry and chemical biology of H_2S, hydropersulfides, and derived species: implications of their possible biological activity and utility. Free Radic Biol Med 77:82–94. https://doi.org/10.1016/j.freeradbiomed.2014.09.007

11. Jung M, Kasamatsu S, Matsunaga T, Akashi S, Ono K, Nishimura A, Morita M, Abdul Hamid H, Fujii S, Kitamura H, Sawa T, Ida T, Motohashi H, Akaike T (2016) Protein polysulfidation-dependent persulfide dioxygenase activity of ethylmalonic encephalopathy

protein 1. Biochem Biophys Res Commun 480 (2):180–186. https://doi.org/10.1016/j.bbrc.2016.10.022

12. Paul BD, Snyder SH (2012) H_2S signalling through protein sulfhydration and beyond. Nat Rev Mol Cell Biol 13(8):499–507. https://doi.org/10.1038/nrm3391

13. Wedmann R, Onderka C, Wei S, Szijarto IA, Miljkovic JL, Mitrovic A, Lange M, Savitsky S, Yadav PK, Torregrossa R, Harrer EG, Harrer T, Ishii I, Gollasch M, Wood ME, Galardon E, Xian M, Whiteman M, Banerjee R, Filipovic MR (2016) Improved tag-switch method reveals that thioredoxin acts as depersulfidase and controls the intracellular levels of protein persulfidation. Chem Sci 7(5):3414–3426. https://doi.org/10.1039/c5sc04818d

14. Greiner R, Palinkas Z, Basell K, Becher D, Antelmann H, Nagy P, Dick TP (2013) Polysulfides link H_2S to protein thiol oxidation. Antioxid Redox Signaling 19 (15):1749–1765. https://doi.org/10.1089/ars.2012.5041

15. Nagy P, Winterbourn CC (2010) Rapid reaction of hydrogen sulfide with the neutrophil oxidant hypochlorous acid to generate polysulfides. Chem Res Toxicol 23(10):1541–1543. https://doi.org/10.1021/tx100266a

16. Nagy P, Palinkas Z, Nagy A, Budai B, Toth I, Vasas A (2014) Chemical aspects of hydrogen sulfide measurements in physiological samples. Biochim Biophys Acta 1840(2):876–891. https://doi.org/10.1016/j.bbagen.2013.05.037

17. Vasas A, Doka E, Fabian I, Nagy P (2015) Kinetic and thermodynamic studies on the disulfide-bond reducing potential of hydrogen sulfide. Nitric Oxide 46:93–101. https://doi.org/10.1016/j.niox.2014.12.003

18. Shen X, Peter EA, Bir S, Wang R, Kevil CG (2012) Analytical measurement of discrete hydrogen sulfide pools in biological specimens. Free Radic Biol Med 52(11–12):2276–2283. https://doi.org/10.1016/j.freeradbiomed.2012.04.007

19. Wintner EA, Deckwerth TL, Langston W, Bengtsson A, Leviten D, Hill P, Insko MA, Dumpit R, VandenEkart E, Toombs CF, Szabo C (2010) A monobromobimane-based assay to measure the pharmacokinetic profile of reactive sulphide species in blood. Br J Pharmacol 160(4):941–957. https://doi.org/10.1111/j.1476-5381.2010.00704.x

20. Rabilloud T, Vuillard L, Gilly C, Lawrence JJ (1994) Silver-staining of proteins in polyacrylamide gels: a general overview. Cell Mol Biol 40(1):57–75

21. Havlis J, Thomas H, Sebela M, Shevchenko A (2003) Fast-response proteomics by accelerated in-gel digestion of proteins. Anal Chem 75(6):1300–1306

22. Shevchenko A, Wilm M, Vorm O, Mann M (1996) Mass spectrometric sequencing of proteins silver-stained polyacrylamide gels. Anal Chem 68(5):850–858

23. Eriksson S, Prigge JR, Talago EA, Arner ES, Schmidt EE (2015) Dietary methionine can sustain cytosolic redox homeostasis in the mouse liver. Nat Commun 6:6479. https://doi.org/10.1038/ncomms7479

24. Longen S, Richter F, Kohler Y, Wittig I, Beck KF, Pfeilschifter J (2016) Quantitative persulfide site identification (qPerS-SID) reveals protein targets of H2S releasing donors in mammalian cells. Sci Rep 6:29808. https://doi.org/10.1038/srep29808

25. Alvarez B, Carballal S, Turell L, Radi R (2010) Formation and reactions of sulfenic acid in human serum albumin. Methods Enzymol 473:117–136. https://doi.org/10.1016/S0076-6879(10)73005-6

26. Miseta A, Csutora P (2000) Relationship between the occurrence of cysteine in proteins and the complexity of organisms. Mol Biol Evol 17(8):1232–1239

27. Holmberg A, Blomstergren A, Nord O, Lukacs M, Lundeberg J, Uhlen M (2005) The biotin-streptavidin interaction can be reversibly broken using water at elevated temperatures. Electrophoresis 26(3):501–510. https://doi.org/10.1002/elps.200410070

Chapter 6

Vascular Effects of H$_2$S-Donors: Fluorimetric Detection of H$_2$S Generation and Ion Channel Activation in Human Aortic Smooth Muscle Cells

Alma Martelli, Valentina Citi, and Vincenzo Calderone

Abstract

Hydrogen sulfide (H$_2$S) evokes vascular effects through several mechanisms including in wide part the activation of some ion channels such as ATP-sensitive potassium (K$_{ATP}$) channels and voltage-gated Kv7 potassium channels. Electrophysiological methods are very accurate, but they require high expertise and high specialized equipment. A more manageable fluorimetric technique which allows to record the membrane potential variations by the employment of an anionic bis-oxonol dye named DiBac4(3) with the administration of different blockers of several potassium channels could be useful to discover the targets of H$_2$S-induced vascular hyperpolarization. Coupled with this technique, a fluorimetric detection (by the use of WSP-1 dye) of H$_2$S generation in human vascular smooth muscle cells after H$_2$S-donor administration could confirm the ability of these molecules to evoke the hyperpolarizing effect through the H$_2$S release.

Key words Human aortic smooth muscle cells (HASMCs), DiBac4(3), WSP-1, Hydrogen sulfide detection, Hyperpolarizing effect, Potassium channels

1 Introduction

The gaseous transmitter, hydrogen sulfide (H$_2$S), was able to evoke vascular effects both in vitro and in vivo, and among the several mechanisms of action proposed, the activation of some types of potassium channels as ATP-sensitive potassium (K$_{ATP}$) channels and Kv7 potassium channels and consequent vascular smooth muscle (VSM) hyperpolarization were the most accredited and investigated [1–3]. Certainly, the direct electrophysiological techniques, such as patch clamp method, represent the gold standard for studies on ion channels, but unfortunately they require solid expertise of the operator and high specialized lab equipment not so easy to find. However the effects of H$_2$S on the membrane potential of cultured human aortic smooth muscle cells (HASMCs) could be tested by means of the membrane potential-sensitive fluorescent dye bis

(1,3-dibutylbarbituric acid)trimethineoxonol (DiBac4(3)), allowing an indirect electrophysiological recording [3–5]. In fact, DiBac4(3) distributes itself between the membrane according the Nernst potential and determines a decrease in fluorescence when it flows outside cells following a hyperpolarization; conversely, it determines an increase in fluorescence when it flows inside cells following a depolarizing stimuli [6, 7]. This technique, through the application of several blockers of the different types of potassium channels, allows to isolate the current, and so the channel, responsible of the hyperpolarizing effect on HASMC after the administration of H_2S-donors [8]. Coupled with this technique, it is fundamental to detect H_2S generated by the application of H_2S-donors on HASMC, and this recording is possible through the use of another fluorimetric dye named WSP-1 [9]. WSP-1 is a H_2S-specific fluorescent dye able to react with H_2S, releasing a fluorophore [10]. These techniques taken together give a lot of information about the H_2S-donor properties and the associated hyperpolarizing vascular effects of some novel compounds designed to be new molecular entities able to release H_2S.

2 Materials

1. Human aortic smooth muscle cell (HASMC).

2. Medium 231 supplemented with smooth muscle growth supplement (SMGS) and 1% of 100 units/mL penicillin and 100 mg/mL streptomycin.

3. Gelatin 1% (from porcine skin): weigh 1 g of gelatin from porcine skin in a 250 mL glass beaker. Add 100 mL of sterile Dulbecco's phosphate-buffered saline (DPBS). Mix gently to solubilize gelatin by a heating magnetic stirrer at about 80 °C. When the solution comes back to room temperature, filter it through a 0.2 μm Corning filter mounted on a 10 mL syringe, aliquot it in 15 mL conical centrifuge tubes, and store aliquots at 4 °C before use (*see* **Note 1**).

4. Buffer standard (BS): add 500 mL of sterile DPBS to a 1 L glass beaker. Weigh 2.383 g of HEPES (20 mM), 3.506 g of NaCl (120 mM), 0.0745 g of KCl (2 mM), 0.148 g of $CaCl_2 \cdot 2H_2O$ (2 mM), 0.1015 g of $MgCl_2 \cdot 6H_2O$ (1 mM), and 0.4955 g of glucose (5 mM); add one by one all the compounds to the 500 mL of DPBS mixing the solution gently by a magnetic stirrer at room temperature. At the end of solubilization, adjust the pH to 7.4.

5. 96-well black plates: clear bottom pre-coated with gelatin 1%.

2.1 Vascular Smooth Muscle Membrane Hyperpolarization by H₂S-Donors

1. DiBac4(3) stock solution: Add dimethyl sulfoxide (DMSO) to the membrane potential-sensitive fluorescent dye bis (1,3-dibutylbarbituric acid)trimethineoxonol (DiBac4(3)) in order to obtain a 500 μM stock solution, and store it at −20 °C. During the preparation and the storing period, avoid any contact with light except for the infrared (IR) one.

2. DiBac4(3) working solution: Dilute 75 μL of 500 μM DiBac4 (3) stock solution in 15 mL of BS in order to obtain a 2.5 μM working solution. During the preparation, avoid any contact with light except for the infrared (IR) one.

2.2 Fluorimetric Detection of Intracellular H₂S

1. WSP-1 stock solution: Add DMSO to the fluorescent dye 1, 3′ -methoxy-3-oxo-3H-spiro[isobenzofuran-1,9′-xanthen]-6′yl2 (pyridin-2-yldisulfanyl)benzoate (Washington State Probe-1, WSP-1) in order to obtain a 2 mM stock solution, and store it at −20 °C. During the preparation and the storing period, avoid any contact with light except for the infrared (IR) one.

2. WSP-1 working solution: Dilute 500 μL of 2 mM WSP-1 stock solution in 9.5 mL of BS in order to obtain a 100 μM working solution. During the preparation, avoid any contact with light except for the infrared (IR) one.

3. Bouin solution: 5% acid acetic, 9% formaldehyde, 9% picric acid in distilled water.

3 Methods

All the procedure should be carried out avoiding any contact with light except for the IR one.

3.1 Vascular Smooth Muscle Membrane Hyperpolarization by H₂S-Donors

1. Culture HASMCs up to about 90% confluence, and 24 h before the experiment, seed cells onto a 96-well black plate; clear bottom pre-coated with gelatin 1% (from porcine skin), at density of 72×10^3 per well.

2. After 24 h to allow cell attachment, replace the culture medium, and incubate the cells for 1 h in the previously described DiBac4(3) working solution, 180 μL per well. In a specular manner, plan also to fill with 180 μL of DiBac4 (3) working solution (and following different treatments) a series of wells named "blanks," which will receive the same treatments of tested wells but without cells (*see* **Note 2**). The membrane potential-sensitive dye DiBac4(3) allows a non-electrophysiological measurement of cell membrane potential; in fact, this lipophilic and negatively charged oxonol dye shuffles between cellular and extracellular fluids in a

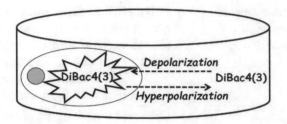

Fig. 1 Schematic representation of DiBac4(3) mechanism of action inside and outside cells. DiBac4(3) is an anionic dye able to give fluorescence only when it binds intracellular proteins. During incubation into the well, DiBac4(3) distributes itself between the intracellular and the extracellular sides according to the Nernst potential. After the administration of a hyperpolarizing agent, DiBac4 (3) flows outside cells giving a decrease in fluorescence; on the contrary, after the administration of a depolarizing stimuli, DiBac4(3) flows inside cells giving an increase of fluorescent signal

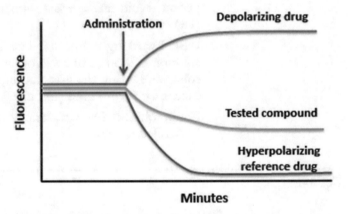

Fig. 2 Schematic representation of typical track obtained using DiBac4 (3) method. After the administration of the tested compounds or of the reference drugs, the trends of fluorescence are followed for a period of about 30–40 min, and the fluorescence is recorded every 2.5 min

membrane potential-dependent manner (following the Nernst law), thus allowing to assess changes in membrane potential by means of spectrofluorimetric recording (Fig. 1). In particular, an increase of fluorescence, corresponding to an inward flow of the dye, reflects a membrane depolarization; in contrast, a decrease in fluorescence, due to an outward flow of the dye, is linked to membrane hyperpolarization (Fig. 2).

3. At the end of DiBac4(3) incubation, add 10 μL of several blockers of different types of ion channels (such as glybenclamide for K_{ATP} channels, XE991 for Kv7 channels, iberiotoxin for BK channels, and so on), and incubate them for 20 min. Every blocker should be diluted in an appropriate vehicle and then diluted to reach a minimal percentage of this vehicle in the

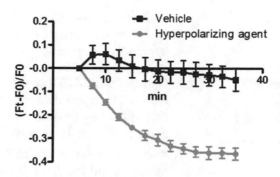

Fig. 3 Typical experimental track. This typical track was obtained recording the variation of fluorescence every 2.5 min for a period of 35 min. The line with black squares represent the physiological low decrease of fluorescence in the presence of the corresponding vehicle used to dissolve the tested hyperpolarizing agent (such as DMSO for H$_2$S-donors). The line with gray-fill circles represents the significant decrease of fluorescence obtained after the administration of a hyperpolarizing agent like a H$_2$S-donor

final concentration at contact with cells (e.g., DMSO $\leq 0.5\%$ on HASMC). For every blocker treatment, a triplet of wells treated only with the corresponding vehicle at the same dilution percentage must be planned.

4. At the end of the 20-min incubation with the respective blocker, read the basal fluorescence of every well through a multiplate reader, on the spectrofluorometer function at excitation and emission wavelengths of 488 and 520 nm, respectively (*see* **Note 3**).

5. After the assessment of baseline fluorescence, add 10 μL of the tested H$_2$S-donor hyperpolarizing compounds at the desired concentration. Every H$_2$S-donor tested compound should be diluted in an appropriate vehicle and then diluted to reach a minimal percentage of this vehicle in the final concentration at contact with cells (e.g., DMSO $\leq 0.5\%$ on HASMC). For every tested compound treatment, a triplet of wells treated only with the corresponding vehicle at the same dilution percentage must be planned (*see* **Notes 4** and **5**).

6. Follow the trends of fluorescence for 40 min. Record the relative fluorescence decrease, linked to hyperpolarizing effects, every 2.5 min, and then calculate it as:

$$(F_t - F_0)/F_0$$

where F_0 is the basal fluorescence before the addition of the tested compounds and F_t is the fluorescence at time t after their administration (Fig. 3). Subtract the fluorescence recorded values of the corresponding "blank wells" from the fluorescence values recorded for every treated well. Calculate the area under the curve (AUC) of time-fluorescence relationship, and express it as a percent of that induced by the selected reference hyperpolarizing drug.

3.2 Fluorimetric Detection of Intracellular H₂S

3.2.1 Spectrofluorimetric Assessment

1. Culture HASMCs up to about 90% confluence, and 24 h before the experiment, seed cells onto a 96-well black plate; clear bottom pre-coated with gelatin 1% (from porcine skin), at density of 30×10^3 per well.

2. After 24 h to allow cell attachment, replace the culture medium, and incubate the cells for 30 min at 37 °C avoiding light exposure in the previously described WSP-1 working solution (allowing cells to upload the dye), 200 μL per well. In a specular manner, plan also to fill with 200 μL of WSP-1 working solution (and following different treatments) a series of wells named "blanks," which will receive the same treatments of tested wells but without cells.

3. After 30 min of incubation, replace the WSP-1 working solution with 190 μL of buffer standard (both for the wells seeded with cells and for wells named "blanks"). This step allows to decrease and minimize the interference of the extracellular WSP-1 when the H₂S-donor is added.

4. Read the fluorescence signal with a spectrofluorometer at $\lambda = 465/515$ nm. This value represents the basal fluorescence due to intracellular endogenous H₂S.

5. Add 10 μL of the tested H₂S-donor compounds at the desired concentration. Every H₂S-donor tested compound should be diluted in an appropriate vehicle and then diluted to reach a minimal percentage of this vehicle in the final concentration at contact with cells (e.g., DMSO ≤0.5% on HASMC). For every tested compound treatment, a triplet of wells treated only with the corresponding vehicle at the same dilution percentage must be planned.

6. When WSP-1 reacts with H₂S generated from H₂S-donors, it releases a fluorophore (Fig. 4) detectable with a spectrofluorometer at $\lambda = 465/515$ nm. Monitor the increase in fluorescence (expressed as fluorescence index = FI) for 1 h reading the fluorescence value every 5 min (Fig. 5). This increase is linked to the H₂S release from the tested compound.

7. For the analysis of the data, subtract the fluorescence values recorded in the "blank wells" from fluorescence values recorded for every treated well. Calculate the area under the curve (AUC) of time-fluorescence relationship which represents the amount of H₂S released by the tested compound in the intracellular side.

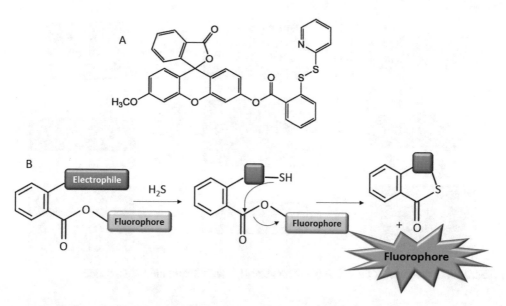

Fig. 4 (**a**) Structure of Washington State Probe-1 (WSP-1). (**b**) Reaction of WSP-1 with hydrogen sulfide, leading to fluorophore release. WSP-1 selectively and rapidly reacts with H₂S to generate benzodithiolone and a fluorophore with excitation and emission of 465 and 515 nm, respectively

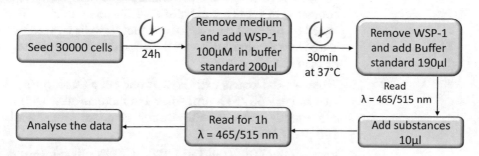

Fig. 5 Schematic diagram for spectrofluorimetric detection of intracellular H₂S

3.2.2 Fluorimetric Microscope Assessment

1. Culture HASMCs up to about 90% confluence, and 24 h before the experiment, seed cells onto a 8-well culture slide; clear bottom pre-coated with gelatin 1% (from porcine skin), at density of 30×10^3 per well.

2. After 24 h to allow cell attachment, replace the culture medium, and incubate the cells for 30 min at 37 °C in the previously described WSP-1 working solution (allowing cells to upload the dye), 400 μL per well.

3. After 30 min of incubation, replace the WSP-1 working solution with 380 μL of buffer standard. This step allows to decrease and minimize the interference of the extracellular WSP-1 when the H₂S-donor is added.

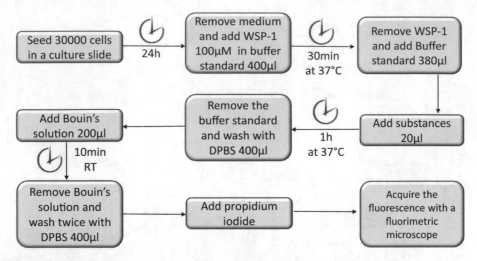

Fig. 6 Schematic diagram for detection of intracellular H_2S with fluorimetric microscope

4. Add 20 μL of the tested H_2S-donor compounds at the desired concentration. Every H_2S-donor tested compound should be diluted in an appropriate vehicle and then diluted to reach a minimal percentage of this vehicle in the final concentration at contact with cells (e.g., DMSO ≤0.5% on HASMC). For every tested compound treatment, a triplet of wells treated only with the corresponding vehicle at the same dilution percentage must be planned.

5. When WSP-1 reacts with H_2S, it releases a fluorophore detectable at $\lambda = 465/515$ nm. After 1-h incubation with the tested compounds, remove the buffer standard, and wash with 400 μL of DPBS.

6. Remove the DPBS, and add 200 μL of Bouin solution per well for 10 min at room temperature to fix the cells.

7. Remove the excess of Bouin solution and wash the cells twice with 400 μL of DPBS.

8. Finally, add propidium iodide to highlight label nuclei. Then, mount the culture slide in order to examine and acquire the fluorescence signal with a fluorescence microscope, and process it with a graphic program (Fig. 6).

4   Notes

1. In preparing gelatin, avoid the formation of bubbles, due to boiling, on the surface of the solution, and, when you filter it, change the filter every 30 mL to avoid the stoppage of the flow.

2. When you replace the culture medium with fluorescent dye working solution, remove the medium gently with a pipette to avoid the removal of the seeded cells: HASMCs are very sensitive to every mechanical trouble.

3. To allow cells work at right temperature during the experiment, set the multiplate reader temperature at 37 °C when you record the fluorescence values for a long period (e.g., 40 min or 1 h) every 2.5 or 5 min.

4. Add the tested compounds (e.g., H₂S-donors) first in the wells named "blank" and then into wells seeded with cells to avoid the loss of the first-minute effect of H₂S-donors on cells.

5. To avoid a loss of H₂S, prepare every solution, especially the tested H₂S-donor solutions, freshly, few minutes before the addition in wells seeded with the cells.

References

1. Martelli A, Testai L, Breschi MC, Blandizzi C, Virdis A, Taddei S, Calderone V (2012) Hydrogen sulphide: novel opportunity for drug discovery. Med Res Rev 32:1093–1130

2. Martelli A, Testai L, Marino A, Breschi MC, Da Settimo F, Calderone V (2012) Hydrogen sulphide: biopharmacological roles in the cardiovascular system and pharmaceutical perspectives. Curr Med Chem 19:3325–3336

3. Martelli A, Testai L, Breschi MC, Lawson K, McKay NG, Miceli F, Taglialatela M, Calderone V (2013) Vasorelaxation by hydrogen sulphide involves activation of Kv7 potassium channels. Pharmacol Res 70:27–34

4. Martelli A, Testai L, Citi V, Marino A, Pugliesi I, Barresi E, Nesi G, Rapposelli S, Taliani S, Da Settimo F, Breschi MC, Calderone V (2013) Arylthioamides as H₂S donors: L-cysteine-activated releasing properties and vascular effects in vitro and in vivo. ACS Med Chem Lett 4:904–908

5. Martelli A, Testai L, Citi V, Marino A, Bellagambi FG, Ghimenti S, Breschi MC, Calderone V (2014) Pharmacological characterization of the vascular effects of aryl isothiocyanates: is hydrogen sulfide the real player? Vasc Pharmacol 60:32–41

6. Gopalakrishnan M, Whiteaker KL, Molinari EJ, Davis-Taber R, Scott VE, Shieh CC, Buckner SA, Milicic I, Cain JC, Postl S, Sullivan JP, Brioni JD (1999) Characterization of the ATP-sensitive potassium channels (K_{ATP}) expressed in guinea pig bladder smooth muscle cells. J Pharmacol Exp Ther 289:551–558

7. Miller TR, Taber RD, Molinari EJ, Whiteaker KL, Monteggia LM, Scott VE, Brioni JD, Sullivan JP, Gopalakrishnan M (1999) Pharmacological and molecular characterization of ATP-sensitive K^+ channels in the TE671 human medulloblastoma cell line. Eur J Pharmacol 370:179–185

8. Martelli A, Manfroni G, Sabbatini P, Barreca ML, Testai L, Novelli M, Sabatini S, Massari S, Tabarrini O, Masiello P, Calderone V, Cecchetti V (2013) 1,4-Benzothiazine ATP-sensitive potassium channel openers: modifications at the C-2 and C-6 positions. J Med Chem 56:4718–4728

9. Liu C, Pan j LS, Zhao Y, Wu LY, Berkman CE, Whorton AR, Xian M (2011) Capture and visualization of hydrogen sulfide via a fluorescent probe. Angew Chem Int Ed Engl 50:10327–10329

10. Martelli A (2016) Vascular effects of p-carboxyphenyl-isothiocyanate, a novel H₂S-donor. In: Abstracts book of the 4th international conference on the biology of hydrogen sulfide, Naples, 3–5 June 2016

Chapter 7

In Vitro Measurement of H_2S-Mediated Vasoactive Responses

Sona Cacanyiova and Andrea Berenyiova

Abstract

Isolated tissue chamber bath system and wire myograph were developed for "in vitro" investigation of vasoactive responses on isolated arteries from a variety of animal species and vascular beds. The chapter characterizes the main principles of mechanical measurement of the changes in isometric tension of vascular smooth muscles in isolated rat thoracic aorta and superior mesenteric artery and describes several protocols on how to investigate vasoactive properties of hydrogen sulfide (H_2S) from the point of view of its mutual interaction with NO. Several methodological advances, results, and their interpretations in the context of the general knowledge are described. In the protocols the approach on how to study the vasoactive modulatory as well as direct action of H_2S and mutual interaction of H_2S with nitroso compounds, lipids, and endogenously produced NO is described.

Key words Isolated artery, Vasoactivity, NaHS, Na_2S, Nitrosothiol, NO synthase, Wistar rat, SHR

1 Introduction

Isolated tissue chamber bath system and wire myograph were developed for "in vitro" investigation of vasoactive responses on isolated arteries from a variety of animal species and vascular beds. The main principle of both systems consists in the measurements of the changes in isometric tension of vascular smooth muscle cells which enables the investigation of both active and passive properties of the isolated arteries. This system represents the method for investigation of vasoactive responses of vascular smooth muscles: vasoconstriction and vasorelaxation. Whereas the multichamber organ bath system is used for investigation of vasoactive properties of large conduit arteries, a wire myograph is a device used for smaller vessels (diameter between 100 and 1000 μm). Using this direct measurement of the mechanics of isolated vascular samples is preferable to different indirect measurements. Rat thoracic aorta (large conduit artery) and superior mesenteric artery (smaller

Jerzy Bełtowski (ed.), *Vascular Effects of Hydrogen Sulfide: Methods and Protocols*, Methods in Molecular Biology, vol. 2007, https://doi.org/10.1007/978-1-4939-9528-8_7, © Springer Science+Business Media, LLC, part of Springer Nature 2019

conduit artery) were used as samples for demonstration of the measurement.

One of the most important mechanisms of H_2S signalling pathways is its involvement in cross talk with nitric oxide (NO) signalization. NO, which is synthesized by three isoforms of NO synthase, binds to the thiol group of different thiols, such as glutathione, cysteine, and albumin, altering their function. Endogenous nitrosothiols may act as intermediates in the storage and/or transport of NO to places where it is utilized in smooth muscle cells and may serve as stores and carriers of NO as a part of the nitroso signalling pathway [1]. Analogous to NO and nitroso-signalization, the synthesized H_2S can be released from cells to affect tissues in a paracrine fashion, or H_2S can initially be stored as sulfane sulfur, which is a divalent sulfur molecule bound to another sulfur molecule, such as outer sulfur atoms of persulfides and inner chain atoms of polysulfides [2]. Thus, as a part of the nitroso signalling pathway, nitroso compounds serve as stores and carriers of NO; as a part of the sulfide signalling pathway, bound sulfane sulfur compounds serve as stores and carriers of H_2S. According to these facts, protein-bound H_2S (or more precisely, protein-bound sulfur) can modulate NO releasing from endogenous NO donors to act in situ. The aim of this chapter is to characterize the main principles of mechanical measurement by abovementioned devices and to describe several protocols on how to investigate vasoactive properties of H_2S from the point of view of its mutual interaction with NO. We describe several methodological advances, results, and their interpretations in the context of the general knowledge.

2 Materials

A laboratory rat is an experimental animal model used for isolation of the different types of arteries and investigation of their vasoactive properties. The Wistar rat is an outbred albino normotensive rat developed for use in biological and medical research. For evaluation of the changes in vasoactivity associated with the development of essential arterial hypertension, a model of spontaneously hypertensive rat (SHR) is used. Like in human beings, the hypertensive response starts with advancing age in this strain of rats, and the cause of the rising blood pressure remains unknown. In our experiments, rats are cared for, and all procedures are performed in accordance with institutional guidelines and are approved by the State Veterinary and Food Administration of the Slovak Republic and by an ethical committee according to the European Convention for the Protection of Vertebrate Animals Used for Experimental and Other Scientific Purposes, Directive 2010/63/EU of the European Parliament. All rats are housed under a 12-h light to 12-h dark cycle, at a constant humidity (45–65%) and temperature (20–22 °C), with free access to standard laboratory rat chow and drinking water. The Institute of Normal and Pathological Physiology provides veterinary care.

The frequently used large conduit artery in in vitro vasoactive studies is thoracic aorta. The aorta is the ultimate artery divided by the diaphragm into the thoracic and abdominal aorta. The aortic wall is composed histologically of three layers: a thin inner tunica intima lined by the endothelium; a thick tunica media characterized by concentric sheets of elastic and collagen fibers with the border zone of the lamina elastica interna and externa, as well as smooth muscle cells; and the outer tunica adventitia containing mainly collagen, vasa vasorum, and the vessel deposits of the fat—perivascular adipose tissue. The aorta plays, besides the conduit function, also an important role in the control of systemic vascular resistance and heart rate, and through its elasticity, it plays an important role in the control of blood pressure.

The superior mesenteric artery arises from the anterior surface of the abdominal aorta and takes and distributes blood to a large portion of the gastrointestinal tract. It is also composed of three layers—tunica intima/endothelium, tunica media, and tunica adventitia. As a muscular type of the artery, it has several layers of smooth muscle cells, and adventitia is richly equipped by adrenergic nerve endings. Among the blood vessels studied, the rat superior mesenteric artery offers a very interesting model for the analysis of the vasoactive responses mediated by noradrenaline released from adrenergic nerve endings after transmural nerve stimulation (*see* **Note 1**).

2.1 Used Solutions, Devices, and Softwares

1. Krebs solution: 118 mM NaCl, 5 mM KCl, 25 mM NaHCO$_3$, 1.2 mM MgSO$_4$ × 7H$_2$O, 2.2 mM KH$_2$PO$_4$, 2.5 mM CaCl$_2$ × 2H$_2$O, 11 mM glucose, 0.032 mM CaNa$_2$EDTA.

2. KPSS (modified Krebs solution): 123.7 mM KCl, 25 mM NaHCO$_3$, 1.17 mM MgSO$_4$ × 7H$_2$O, 1.18 mM KH$_2$PO$_4$, 2.5 mM CaCl$_2$ × 2H$_2$O, 5.5 mM glucose, 0.03 mM Na$_2$EDTA × 2H$_2$O.

3. NaHS stock solution: 100 mM NaHS in 160 mM KCl containing 1 mM MgCl$_2$, 0.1 mM diethylenetriaminepentaacetic acid (DTPA), 10 mM HEPES, and 5 mM Tris–Hcl, pH 7.4. Prepare only at the time of measurement and use within a few hours. NaHS dissociates in solution to Na$^+$ and HS$^-$ and after reaction with H$^+$ yields H$_2$S; use the term "sulfide" to encompass the total mixture of H$_2$S, HS$^-$, and S^{2-}.

4. Na$_2$S (Na$_2$S × 9 H$_2$O) stock solution: 100 mM Na$_2$S in ultrapure deionized water (≥18 MΩ.cm) (Millipore, Darmstadt, Germany). Aliquot stock solutions in plastic vials, and keep frozen at −80 °C. At the day of the experiment, thaw 100 μL

of the stock solution, and dilute with 200 mM Tris–HCl (pH 7.4). Prepare fresh working Na$_2$S solution before the experiment, keep in sealed vials with minimal headspace, and use immediately (*see* **Note 2**). Na$_2$S dissociates in solution and reacts with H$^+$ to yield HS$^-$, H$_2$S, and a trace of S^{2-}; use the term "sulfide" to encompass the total mixture of H$_2$S, HS$^-$, and S^{2-}.

5. S-Nitrosoglutathione (GSNO): 10 mM GSNO in 200 mM Tris–HCl containing 0.1 mM DTPA, pH 7.4. Aliquot stock solution in plastic vials and put into −80 °C freezer until use. Check the concentration of stock solution spectrophotometrically using extinction coefficient of 922 M^{-1} cm^{-1} at 335 nm. At the day of the experiment, thaw the stock solution, and dilute with 200 mM Tris–HCl, pH 7.4. Keep in sealed vials, and use within 2 h (*see* **Notes 2** and **3**).

6. Noradrenaline (NA) stock solution: 1 μM NA in distilled deionized water. Prepare the solution fresh before the experiment and use within the day (*see* **Note 3**).

7. Phenylephrine (Phe) stock solution: 1 μM *R*-(−)-Phenylephrine hydrochloride in deionized water. Keep stock solution at 5 °C. Prepare working solution by diluting stock solution with Krebs solution to appropriate concentration before the experiment, and use within a day.

8. Acetylcholine (Ach) stock solution: 10 μM acetylcholine chloride in deionized water. Keep stock solution at 5 °C. Prepare working solution by diluting stock solution with Krebs solution to appropriate final concentration before the experiment, and use within a day (*see* **Note 3**).

9. Nω-Nitro-ʟ-arginine methyl ester (ʟ-NAME): 1 μM ʟ-NAME in deionized H$_2$O. Keep stock solution at −20 °C. Dilute with Krebs solution to the appropriate final concentration before the experiment, and use within the day.

10. 1*H*-[1,2,4]Oxadiazolo[4,3-a]quinoxalin-1-one   (ODQ): 10 μM ODQ in DMSO. Keep stock solution at 5 °C, and dilute with the Krebs solution before the experiment; use the working solution within a day.

11. (4-Carboxyphenyl)-4,4,5,5-tetramethylimidazoline-1-oxyl-3-oxide potassium salt (cPTIO): 100 μM cPTIO in deionized water. Keep stock solution at 5 °C, dilute with Krebs solution to appropriate concentration before the experiment, and use within a day.

12. Nitrosylated bovine serum albumin (NO-BSA): prepare 1 μM NO-BSA by mixing 2 mM NaNO$_2$ and 2 mM BSA in the dark at a room temperature for 10 min. Bring pH to 7.2 by adding 1 mM NaOH. Yield of BSA-NO is 50 mM (*see* **Note 2**). Store BSA-NO stock solution at −70 °C.

13. Asolectin: 800 μM asolectin in 160 mM KCl containing 1 mM MgCl$_2$, 0.1 mM DTPA, 10 mM HEPES, and 5 mM Tris, pH 7.4. Vortex for 2 min and sonicate for 10 min (Bandelin Sonorex RK-31, 35 kHz, Germany).

14. Na$_2$S-GSNO products: mix 1 mM GSNO and 10 mM Na$_2$S at 21 ± 2 °C to reach a 10:1 molar excess of Na$_2$S over GSNO, and wait 3 min until complete formation of reaction products (*see* **Notes 2–4**). Dilute with 200 mM Tris–HCl, pH 7.4 to appropriate final concentration. The concentration of the mixture (SSNO) is defined as the concentration of GSNO.

15. Electromechanical force transducers: FSG-01/50gr, ELS-14 (Experimetria/MDE Co. Ltd., Budapest, Hungary).

16. DEWESoft software (Dewetron GmbH, Gambrach, Austria).

17. Kymograph software (now, chemicals and devices are mixed. Electrical stimulator: ST-03-04, 410A, DMT (Experimetria/MDE Co. Ltd., Budapest, Hungary)).

3   Methods

3.1 In Vitro Measurement of Vasoactive Responses: Multichamber Organ Bath System (Thoracic Aorta)

The addition of vasoactive compounds into the multichamber organ bath system leads to the decrease (vasorelaxation) or increase (vasoconstriction) of the arterial wall tone. The modulatory effects of certain vasoactive substance (e.g., of H$_2$S donors) are tested as a pre-treatment inducing a minimal change in a developed tension. The rate of relaxation response is expressed as a percentage of the adrenergic-induced contraction: the increase of the vascular tone is set as 100%, and relaxant effect of selected compounds is given relative to the difference of the tension. The extent of the contractile responses and the changes in the increasing active tension are expressed as developed changes in isometric tension (g) (*see* **Note 5**).

1. Sacrifice rats by decapitation after a brief anesthetization with CO$_2$.

2. Isolate the aorta: remove the segment of aorta from the intact aorta beginning 5 mm below the aortic arch and ending before a diaphragm, and place in Krebs solution.

3. Clean the aorta carefully of fat and connective tissue, and cut into 5 mm length rings (*see* **Note 6**). Take care not to damage the endothelium lining vessels (*see* **Note 7**).

4. Fix the rings vertically between the bases of two triangular shaped tungsten wires, and immerse in 20 mL incubation organ bath with Krebs solution oxygenated with 95% O$_2$ and 5% CO$_2$ mixture, and keep at 37 °C. Fix the bottom triangle onto the hook on the tissue holder and the upper triangle on a glass stick fixed onto the arm of the sensor—electromechanical

force transducer FSG-01/50gr (Experimetria/MDE Co. Ltd., Budapest, Hungary). As the smooth muscle cells of the artery trigger the vasoactive responses, the sensor records the changes in developed isometric tension. The registered signal is amplified and converted by bridge amplifier and AD converter, and the output data are evaluated using DEWESoft or Kymograph software.

5. Place the rings of thoracic aorta under a resting tension of 1 g optimal in control as well as experimental conditions.

6. Maintain the rings throughout 45–60 min of equilibration period until stress relaxation no longer occurs.

7. Check the contractility of the rings by washing out with modified Krebs solution—KPSS.

8. Precontract rings by non-specific agonist of adrenergic receptors noradrenaline (NA, 1 μM) or specific agonist of α_1-adrenergic receptors, phenylephrine (Phe, 1 μM), and after the achievement of stabile plateau, treat the rings with selected compounds (NO donor, H_2S donor, acetylcholine, etc.).

9. Wash out the rings after finishing of the responses three times by Krebs solution in 7-min intervals.

3.2 Vasomodulatory Effect of H₂S Donors

To follow the modulatory effects of H_2S, the low doses of H_2S donors are used (1–40 μM) (see **Note 8**). The ring of aorta in organ bath is treated with modulatory dose of H_2S donor 2–3 min before adding a compound whose effect is modulated. The approach on how to study the vasoactive modulatory action of H_2S and mutual interaction of H_2S with nitroso compounds and lipids is described. Subheading 3.2.1 represents a functional test whether H_2S is able to release NO from nitrosothiols. The experiments are based on the comparison of vasorelaxant responses induced by exogenously administered nitrosothiol (NO donor) before and after a pre-treatment with H_2S donor. The effects of NaHS as a H_2S donor and GSNO are described as an example [3]. Nitrosothiols like endogenous GSNO, nitroso-cysteine, or nitroso-serum albumin may act as an intermediate in the storage and/or transport of NO as a part of nitroso signalling. Serum albumin represents the largest fraction of free thiols in circulation. In Subheading 3.2.2, it is tested if H_2S donor could induce a release of NO from serum albumin [4]. In Subheading 3.2.3, it is tested whether lipids can influence biological effect of H_2S, and the effect of unsaturated fatty acid (asolectin) on H_2S-induced contraction of NA-precontracted thoracic aorta followed [5]. As a source of asolectin, a highly purified phospholipid product from soybean mixture of phospholipids containing lecithin, cephalin, inositol phosphatides, and soybean oil is used (about 24% saturated fatty acids, 14% monounsaturated, and 62% polyunsaturated fatty acids).

*3.2.1 H₂S
and S-Nitrosoglutathione
(The Modulatory Effect
of pH and the Soluble
Guanylate Cyclase
Involvement)*

1. Prepare aortic rings for tension measurement as described above (Subheading 3.1, **steps 1–7**).

2. Apply 1 µM noradrenaline, and after the achievement of stabile plateau, apply a single dose of GSNO (0.5 µM) for 4 min (Fig. 1a). Wash it out three times by Krebs solution in 7-min intervals.

3. Apply 1 µM noradrenaline. After the achievement of stabile tension plateau, add NaHS (30 µM), and after 2–3 min incubation with NaHS, apply again GSNO (0.5 µM) (Fig. 1b). Wash it out three times by Krebs solution in 7-min intervals.

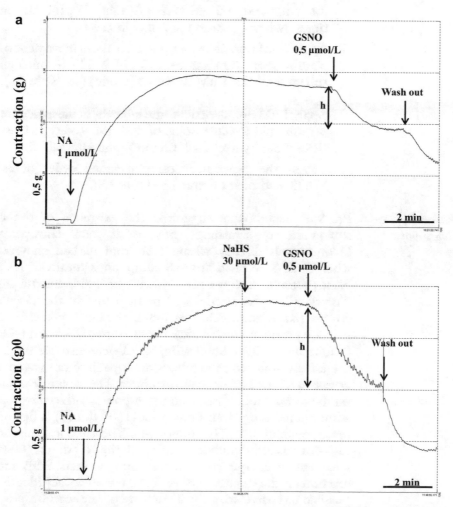

Fig. 1 Representative records of the vasoactive effects of 0.5 µM GSNO on 1 µM noradrenaline-precontracted rings of thoracic aorta before (**a**) and after (**b**) the pre-treatment with H₂S donor. h, extent of relaxation

4. Repeat the same procedure at the pH 6.3. Add 10 mM pipera-zine-N,N'-bis(2-ethanesulfonic acid) (PIPES) and 20 mM $NaHCO_3$ to the Krebs solution.

5. Measure pH directly in the oxygenated solution at 37 °C.

6. Repeat **step 2** (*see* **Note 9**). Wash it out three times by Krebs solution in 7- min intervals.

7. Apply 10 μM ODQ (selective inhibitor of soluble guanylate cyclase, sGC) 15 min before the addition of noradrenaline (*see* **Note 10**).

8. Repeat the same procedure—*see* **step 2**.

3.2.2 H_2S and BSA-NO

1. Apply 1 μM noradrenaline, and after the achievement of stabile plateau, treat the rings by single applications of 1 μM BSA-NO for 4 min and 100 μM NaHS (4 min). Wash them out three times by Krebs solution in 7-min intervals.

2. Apply 1 μM noradrenaline, and after the achievement of stabile plateau, treat the rings by 100 μM NaHS (4 min) and then subsequently by 1 μM BSA-NO (4 min) (*see* **Note 11**).

3.2.3 H_2S and Lipids

1. Apply 1 μM noradrenaline, and after the achievement of stabile plateau, add a single bolus of 800 μM asolectin. Wash it out three times by Krebs solution in 7-min intervals.

2. Repeat the experiment using changed order of asolectin, NaHS, and their mixture (*see* **Note 12**).

3.3 Coupled Sulfide-Nitroso Signalling Pathway

Previous experiments suggested the existence of the coupled sulfide-nitroso signalling, but it is not known whether S-compounds directly release NO from nitroso compounds or whether H_2S released from S-compounds induces NO release. Next protocols uncover the existence of sulfide-nitroso signalling "interface" in the arterial wall participating in the vascular tone control (Subheading 3.3.1) as well as the functional effects of new reaction products created after the mixture of H_2S and NO donors (Subheading 3.3.2). Subheading 3.3.1 demonstrates the ability of the arterial wall "to remember" the capability of exogenous H_2S donor to release NO from nitrosothiols. The experiments are based on the comparison of vasorelaxant responses induced by repeatedly administered exogenous nitrosothiol (NO donor) before and after separate addition of H_2S donor [4]. The chemical interaction of H_2S donor sodium sulfide (Na_2S) with the NO donor (GSNO) has been described to generate new reaction products. While individual reaction product(s) have not yet been unequivocally identified, it is possible to characterize the bioactivity of key reaction products in intact tissue preparations. In Subheading 3.3.2, the vascular effect of the longer-lived products of the Na_2S and GSNO interaction is described [6]. The procedure of Cortese-Krott et al. [7] is used to prepare longer-lived products. Short-lived reaction intermediates can be safely excluded from contributing to the bioactivity of the mixture following this route of preparation.

3.3.1 Arterial Wall "Memory"

1. Apply 1 μM noradrenaline, and after the achievement of stabile plateau, relax the rings by single application of 0.5 μM GSNO for 4 min.

2. Wash out the rings three times by Krebs solutionn 7-min intervals.

3. Repeat **step 1** and **step 2** two times.

4. Apply 1 μM noradrenaline, and after the achievement of stabile plateau, add a single dose of 0.5 μM GSNO for 4 min.

5. Wash out the rings three times by Krebs solution in 7-min intervals.

6. Apply 1 μM noradrenaline, and after the achievement of stabile plateau, apply 100 μM NaHS.

7. Remove NaHS from the organ bath. Wash out the rings three times by Krebs solution in 7-min intervals.

8. Apply 1 μM noradrenaline, and after the achievement of stabile plateau, add a single dose of 0.5 μM GSNO for 4 min (*see* **Note 13**).

3.3.2 Reaction Products of H₂S and GSNO Mixture (Signalling Pathways)

1. Apply 1 μM phenylephrine, and after the achievement of stabile plateau, add the single bolus of 0.1 μM GSNO. Wash it out three times by Krebs solution in 7-min intervals.

2. Apply 1 μM phenylephrine, and after the achievement of stabile plateau, add the mixture of 1 μM Na₂S and 0.1 μM GSNO. Wash it out three times by Krebs solution in 7-min intervals.

3. To study the involvement of sGC in the effects of GSNO and Na₂S-GSNO products, apply 10 μM ODQ 15 min before the addition of 1 μM phenylephrine.

4. Apply 1 μM phenylephrine, and after the achievement of stabile plateau, add the mixture of 1 μM Na₂S and 0.1 μM GSNO. Wash it out three times by Krebs solution in 7-min intervals.

5. To evaluate the importance of NO in the relaxation effects, apply 100 μM cPTIO (the NO scavenger) 15 min before the addition of 1 μM phenylephrine.

6. Apply 1 μM phenylephrine, and after the achievement of stabile plateau, add the mixture of 1 μM Na₂S and 0.1 μM GSNO (*see* **Notes 14** and **15**).

3.4 H_2S and Endogenous NO Production

H_2S and NO can also react in the context of their endogenous production. To study the participation of endogenously produced NO, the pre-treatment with inhibitors of NO synthase is used. Several classes of compounds revealed an inhibitory effect on NO synthase. The best known are L-arginine analogues, such as the non-specific inhibitor of NO synthase, N^G-nitro-L-arginine methy-lester (L-NAME). On the other hand, for the activation of endothelial NO synthase, acetylcholine can be used in functional studies since it has been shown as an activator of NO production in conduit arteries. The following protocols are aimed to study the effect of H_2S on vasorelaxation mediated by NO synthase-produced NO (the effect of the pre-treatment with Na_2S on acetylcholine-induced relaxation) and the effect of the inhibition of endogenously produced NO on vasoactive responses induced by H_2S (the effect of the pre-treatment with L-NAME on Na_2S-induced responses). Moreover, this type of protocol enables to study the direct vasoactive effects of exogenous H_2S.

3.4.1 H_2S and Acetylcholine-Induced Relaxation

1. Apply 1 μM noradrenaline, and after the achievement of stabile plateau, add a single maximal dose (10 μM) of acetylcholine for 4 min.

2. Wash it out three times by Krebs solution in 7-min intervals.

3. Apply 1 μM noradrenaline, and after the achievement of stabile plateau, add 40 μM of NaHS and after 3 min 10 μM acetylcholine (*see* **Note 16**).

3.4.2 H_2S and L-NAME-Induced NO Synthase Inhibition

1. Apply 1 μM noradrenaline, and after the achievement of stabile plateau, administer cumulatively increasing doses of Na_2S (20, 40, 80, 100, and 200 μM). Wash it out three times by Krebs solution in 7-min intervals.

2. Add 1 μM L-NAME into the organ bath for 15 min.

3. After 15 min, apply again 1 μM noradrenaline, and after the achievement of stabile plateau, add Na_2S in a cumulative manner (20, 40, 80, 100, and 200 μM) (*see* **Note 17**).

3.5 In Vitro Measurement of Vasoactive Responses: Wire Myograph (Mesenteric Artery)

3.5.1 Principle of Mechanical Measurement

1. Remove the main branch mesenteric artery (arteria mesenterica superior), and place it in Krebs solution.

2. Clean the artery carefully of fat and connective tissue, and cut into 3 mm length rings. Take care not to damage the endothelium lining vessels.

3. Immerse the artery in 10 mL incubation bath with Krebs solution oxygenated with 95% O_2 and 5% CO_2 mixture, and keep at 37 °C.

4. Pass two thin tungsten wires (diameter 100 μm) through the lumen of the artery.

5. Stretch each wire between two bridges of supporting plate. One plate is connected to the strain gauge, the other to the micrometer control which controls vertical movement. Both bridges are equipped by the electrodes closely lining the surface of the artery horizontally. The electrodes are connected to the electrical stimulator (ST-03-04, 410A, DMT, Experimetria/MDE Co. Ltd. from 2015/Budapest, Hungary).

6. Tighten the wires to the minimum: they are situated closely and parallel to each other in the horizontal plane.

7. Place the vessel under resting tension 1 g. As the smooth muscle cells of the artery trigger the vasoactive responses, the electromechanical transducer (ELS-14, Experimetria/MDE Co. Ltd. from 2015/Budapest, Hungary) records the changes in developed isometric tension. The registered signal is amplified and converted by bridge amplifier and AD converter, and the output data are evaluated using DEWESoft software or Kymograph software.

8. Allow tissue to equilibrate and maintain throughout 45–60 min of equilibration period until stress relaxation no longer occurs and artery is prepared for the measurement of vasoactive responses.

9. Using the electric stimulation, stimulate the intramural nerves: square pulses of supramaximal intensity (>35 V), 0.2 ms pulse width, 100 μs delay, 31.3 ms pulse period, and 32 Hz frequency for a period 20 s. These parameters ensure the stimulation of the adrenergic nerve endings located in the arterial wall.

3.5.2 Experimental Protocol: H₂S and Transmural Nerve Stimulation

1. Check the contractility of the rings by washing out with modified Krebs solution—KPSS—for 2 min, and wash out three times by Krebs solution.

2. Expose the arterial ring to the electrical stimulation (32 Hz) for 20 s. It induces the release of endogenous noradrenaline from adrenergic nerve endings leading to the vasoconstriction.

3. After returning of the vessel tone to the basal value (1 g), repeat the stimulation (5–7 times).

4. Wash out the ring three times by Krebs solution in 7-min intervals.

5. Add 40 μM NaHS to the incubation medium 2–3 min before the induction of the stimulation.

6. Expose the ring to the electrical stimulation as described above.

7. Realize the pre-treatment with Na₂S repeatedly before each stimulation (*see* **Note 18**).

4   Notes

1. In several papers, the vasoactivity of superior mesenteric artery is followed using the multichamber organ bath system. In our opinion using of wire myograph is more suitable because of the better and safe manipulation with small artery (endothelium preservation is ensured). Moreover, even if the tissue holders in chamber organ system may be equipped with electrodes for electric stimulation, the electrode could be fixed in such distance from the tissue (insufficiently closely), so it is not possible to stimulate the nerve endings (the smooth muscles are stimulated instead of the nerves).

2. We consider important to check the concentration of several used solutions informatively by UV-Vis spectrophotometry. The concentration of $Na_2S \times H_2O$ stock solution was determined by the absorbance at λ_{max} 232 nm, $\varepsilon = 7700$. The concentration of polysulfides in this stock was <10 μmol/L as judged by UV-Vis spectrophotometry [8]. The product formation of sulfide-GSNO mixture was also followed by UV-Vis spectroscopy (absorbance increase at λ max 412 nm), and reaction was complete within 3 min, by which time the products were diluted to the final concentration in Tris–HCl buffer [6]. The concentration of GSNO stock solution was determined by the absorbance at λ_{max} 334 nm. The concentration of BSA-NO solution was estimated by absorbance at 334 nm using GSNO as a standard [9]. We control these parameters every time before preparing of the stock solutions.

3. Solutions are light sensitive; therefore, they should be kept in the dark.

4. It is important to pay attention that the sulfide-GSNO mixture should become yellow after Na_2S addition. Moreover, the different period of the incubation before application to the organ bath can alter the chemical and physical properties of sulfide-GSNO mixture. It is important to be consistent with the keeping of the same incubation period (in our experiment until 3 min).

5. The single ring preparation is suitable under isometric, but not under isotonic conditions, because of the small isotonic tension developed in response to the contractile agonist [10]. The spiral and the zigzag strips show a discontinuity in the relaxant responses, probably because of the endothelium removal during the preparation. The reproducibility of both contractile and relaxing responses makes using the rings the most suitable method for the study of vasoactive drugs in thoracic aorta and mesenteric artery.

6. Several authors use for the expression of the magnitude of the vasoconstrictor responses of the rings the value of the developed tension (g) per cross-sectional area (CSA). CSA is calculated: weight of the ring (mg)/[length (mm) × density (mg/mm^3)]. The density of vascular smooth muscle is assumed to be 1.05 mg/mm^3. Based on our experiences, there is a problem to standardize the measurement of the weight of the sample. It is measured at the end of the experiment, and the sample is put on the absorbent paper; nevertheless, it is difficult to ensure the same proportion of exhaustion of the fluid. This measurement error misrepresents next calculation. In our opinion, the expression of the magnitude of the vasoconstrictor responses in grams is more exact. In this case, it is very important to standardize and to ensure using of the same length of the ring (5 mm, thoracic aorta; 3 mm, mesenteric artery). We use an adjustable caliper rule for setting of the same length of the rings.

7. In medium- and large-sized blood vessels, where nitric oxide (NO) is a major mediator of vasorelaxation, endothelial denudation leads to the elimination of vasorelaxation [11], and the vasoactive properties of the arterial wall may be affected by changes in the integrity of the endothelium [12]. In our previous study, we confirmed that endothelial denudation as a type of vascular injury is quickly followed by significant alterations in vascular redox equilibrium involving free radical production [13]. This process could importantly affect responses mediated by molecules such NO and hydrogen sulfide. In experiments, the function of the endothelium could be occasionally checked by monitoring responses to acetylcholine.

8. The range of modulatory concentrations used for H₂S donors is usually in the range 1–40 μM. Nevertheless, it is important to do concentration-dependent response induced by H₂S donor in every experiment because the modulatory dose could also be out of this range depending on the type of vessel used, strain and age of the animal, etc. Sometimes it could be different even if the same conditions are ensured but another source and birth of the rats is used.

9. After the decrease of the pH (to 6.3), the stimulating effect of NaHS on GSNO-induced vasorelaxation is significantly inhibited suggesting that HS⁻ rather than H₂S is responsible for NO-releasing effect.

10. After pre-incubation with ODQ, the application of 30 μM NaHS and subsequent addition of 0.5 μM GSNO do not evoke the vasorelaxation, confirming that sGC pathway is involved in this effect.

11. The experiment confirms that exogenous H_2S donor significantly increases relaxation in the presence of BSA-NO, whereas NaHS alone causes contraction rather than relaxation. It indicates that NaHS may release NO from BSA-NO to cause aortic rings relaxation.

12. Asolectin alone does not influence the vascular tone of precontracted thoracic aorta. H_2S donor alone slightly increases the NA-induced contraction. However, the mixture of 100 μM H_2S with 800 μM asolectin significantly increases the contraction with comparison to H_2S donor alone [5]. The mechanism of this effect is not known, but we may speculate that H_2S-asolectin chemical interaction might be involved forming new biologically active substance. Moreover, Tomaskova et al. [5] confirmed that unsaturated fatty acid, linoleic acid, and lipids having unsaturated fatty acid—asolectin, dioleoylphosphocholine, and dioleoylphosphoserine—depressed H_2S-induced NO release from exogenous GSNO. Products of NaHS (H_2S, HS^-, S_2^-) may chemically interact with unsaturated bonds of fatty acids and decrease an effective H_2S concentration for interaction with endogenous nitrosothiols and decrease the release of NO, thus increasing the vascular tone of precontracted artery.

13. The repeated application and removal of NA and GSNO to aortic rings cause a relaxation effect that slightly increases after each application. However, when aortic rings are treated with H_2S donor and then washed out prior to next application of GSNO, the observed relaxation effect is significantly increased [4]. From these results follows that the increased relaxation is due to the enhanced release of NO from GSNO and that sulfur from the H_2S treatment bounds to component(s) of the aortic rings to induce NO release later, after separate GSNO application. According to the results, it is probable that NO and H_2S interact in the tissue to form an unknown complex of nitrosothiols. This complex triggers physiological effect different from the effects of both NO and H_2S donor.

14. In recent years Na_2S is more frequently used H_2S donor than NaHS in this type of experiments. All protocols with NaHS donor may be repeated by using Na_2S as a H_2S donor.

15. The relaxation effect of the SSNO mix is significantly pronounced, faster and shorter lasting when compared with GSNO alone (Fig. 2). The products of Na_2S-GSNO reaction relax precontracted aortic rings with a more than twofold potency compared with GSNO alone. The onset of vasorelaxation of the reaction products is 7–10 times faster compared with GSNO. When aorta is preincubated with ODQ, the vasorelaxation induced by GSNO and/or Na_2S-GSNO products (0.1 μM) is eliminated confirming that sGC pathway is

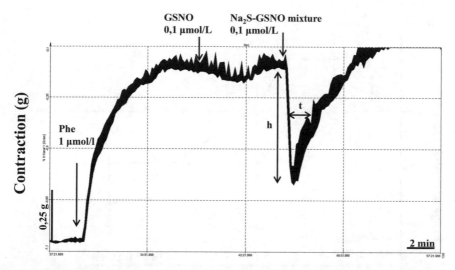

Fig. 2 Representative record of the vasoactive effect of 0.1 μM GSNO and Na₂S-GSNO mixture (0.1 μM) on phenylephrine (1 μM)-precontracted rings of thoracic aorta. h, extent of relaxation; t, the time required for relaxation to return to 50% of the maximum

involved in their effect. The relaxation effect of 0.1 μM GSNO is effectively inhibited by the excess of cPTIO indicating that GSNO acts almost exclusively via releasing NO. On the contrary, relaxation induced by the products of Na₂S-GSNO mixture (0.1 μM) is not inhibited by the presence of cPTIO, indicating that a significant portion of the relaxation induced by the reaction products is mediated by an alternative mechanism, which may directly activate sGC without releasing free NO. Berenyiova et al. [6] confirmed the involvement of more than one product in the reaction mixture accounting for sGC activation and vasorelaxation. While NO is clearly involved in the mechanism of relaxation, nitrosopersulfide (SSNO⁻) and a yet uncharacterized product capable of generating nitroxyl (HNO) appear to account for the pronounced relaxation effects of the Na₂S-GSNO mixture.

16. We performed the described protocol in normotensive Wistar and spontaneously hypertensive rats (SHR). The application of modulator dose of H₂S donor does not affect the acetylcholine-induced vasorelaxation in normotensive rats, but a significant inhibition is observed in SHR [14]. Several studies confirmed that low concentrations of H₂S donors are able to downregulate L-arginine/NO pathway through the inhibition of endothelial NO synthase expression and L-arginine transporter and/or by decreasing NO synthase activity [15, 16]. These effects could be responsible for the inhibition of acetylcholine-induced vasorelaxation after pre-treatment with H₂S donor in SHR suggesting the increased sensitivity of NO-H₂S interaction in hypertensive strain.

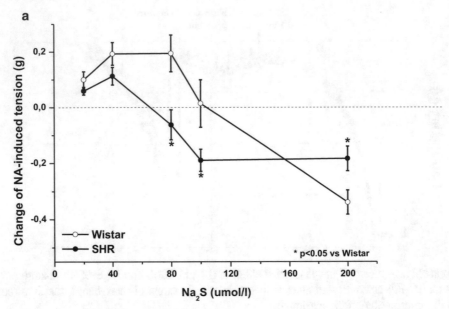

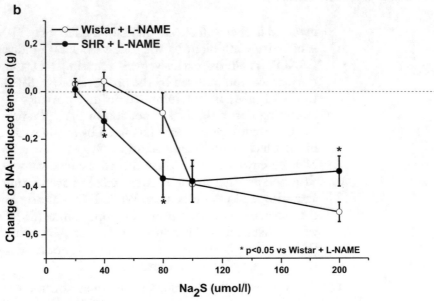

Fig. 3 The concentration-dependent dual effect of Na_2S on noradrenaline (1 μM)-precontracted rings of thoracic aorta (**a**) and its modification after acute endogenous NO inhibition induced by L-NAME (1 μM), (**b**) in Wistar and SHR. Values are mean $\pm$ S.E.M. (*see* **Note 19**)

17. The application of Na_2S on NA-precontracted thoracic aorta rings induces a dual effect in both strains: lower concentrations of Na_2S induce contraction, whereas the higher concentrations evoke vasorelaxation of arterial wall (Fig. 3a). A drift of vasomotor responses in favor of vasorelaxation is observed in SHR compared to Wistar rats because the vasorelaxation is triggered in SHR by lower concentration of Na_2S (80 μM) in comparison

to Wistar rats (200 μM). Moreover, the vasorelaxant effect of higher dose of Na$_2$S (200 μM) is significantly reduced in SHR compared to Wistar rats. An acute inhibition of endogenous NO production modifies the dual effect of Na$_2$S. L-NAME (1 μM) significantly reduces the contractile part of Na$_2$S-induced vasomotor responses in both strains (Fig. 3b). Moreover, the contractile response to Na$_2$S is switched to relaxation at 80 μM in Wistar rats and at 40 μM in SHR suggesting stronger effect and again declaring the increased sensitivity of NO-H$_2$S interaction in SHR in hypertensive strain. This experiment suggests that the acute pre-treatment with L-NAME very probably disabled the inhibitory effect of Na$_2$S on endogenous NO production and/or activity masking the contractile effect of Na$_2$S. Moreover, in young prehypertensive SHR, H$_2$S regulated the arterial tone toward vasorelaxant phase, and this effect was accentuated after the inhibition of endogenous NO. These effects could be a part of the compensatory mechanisms triggered in young SHRs to counter-regulate the increased vascular tone in adulthood [14].

18. The evaluation of the results shows that the application of modulator dose of H$_2$S donor induces significant decrease in adrenergic vasoconstrictor responses induced by transmural nerve stimulation in both strains similarly (Figs. 4 and 5). The mechanism of this effect is not known, but the activity of perivascular adipose tissue could be included. Perivascular adipose tissue is a local deposit of adipose tissue that surrounds the vasculature, and one of the important points in the mediation of its action on vascular function is its interaction with perivascular sympathetic nerves. This tissue is metabolically active and

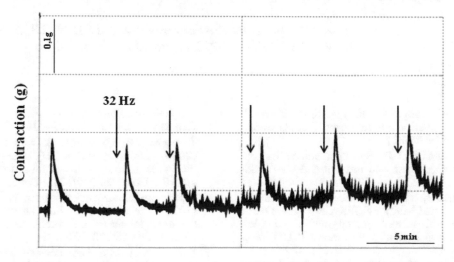

Fig. 4 Representative record of the contractile response of the mesenteric artery induced by transmural nerve stimulation (TNS)

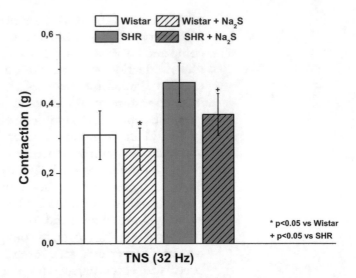

Fig. 5 The modulatory effect of Na_2S pre-treatment on the contractile response induced by transmural nerve stimulation (32 Hz) in Wistar and SHR. Values are mean $\pm$ S.E.M. (*see* **Note 19**)

secretes a wide array of bioactive substances, termed adipo-kines, and H_2S was identified as an adipocyte-derived relaxing factor [17].

19. For the statistical evaluation of differences between groups, a one-way analysis of variance (ANOVA) with a Bonferroni post hoc test was used.

Acknowledgments

Financial support by Slovak grants VEGA 2/0103/18, APVV-15-0565, and APVV-15-0371 is gratefully acknowledged.

References

1. Zhang Y, Hogg N (2005) S-Nitrosothiols: cellular formation and transport. Free Radic Biol Med 38:831–838

2. Kimura H (2010) Hydrogen sulfide: its production, release and functions. Amino Acids 43:113–121

3. Ondrias K, Stasko A, Cacanyiova S et al (2008) H_2S and HS-donor NaHS releases nitric oxide from nitrosothiols, metal nitrosyl complex, brain homogenate and murine L1210 leukaemia cells. Pflugers Arch 457:271–279

4. Bertova A, Cacanyiova S, Kristek F et al (2010) The hypothesis of the main role of H_2S in coupled sulphide-nitroso signalling pathway. Gen Physiol Biophys 29:402–410

5. Tomaskova Z, Cacanyiova S, Benco A et al (2009) Lipids modulate H_2S/HS-induced NO release from S-nitrosoglutathione. Biophys Res Commun 390:1241–1244

6. Berenyiova A, Grman M, Mijusjovic A et al (2015) The reaction products of sulfide and S-nitrosoglutathione are potent vasorelaxants. Nitric Oxide 46:123–130

7. Cortese-Krott MM, Fernandez BO, Santos JL et al (2004) Nitrosopersulfide (SSNO−) accounts for sustained NO bioactivity of

S-nitrosothiols following reaction with sulfide. Redox Biol 2:234–244

8. Nagy P, Pálinkás Z, Nagy A et al (2014) Chemical aspects of hydrogen sulfide measurements in physiological samples. Biochim Biophys Acta 1840:876–891

9. Shi Q, Feng JH, Qu HB et al (2008) A proteomic study of S-nitrosylation in the rat cardiac proteins in vitro. Biol Pharm Bull 31:1536–1540

10. Calderone V, Martinotti E, Scatizzi R et al (1996) A modified aortic multiple-ring preparation for functional studies. J Pharmacol Toxicol Methods 35:131–138

11. Furchgott RF, Zawadzki JV (1980) The obligatory role of endothelial cells in the relaxation of arterial smooth muscle by acetylcholine. Nature 288:373–376

12. Ribback S, Pavlovic D, Herbst D et al (2012) Effects of amitriptyline, fluoxetine, tranylcypromine and venlafaxine on rat vascular smooth muscle in vitro—the role of endothelium. J Physiol Pharmacol 63:119–125

13. Cacanyiova S, Dovinova I, Kristek F (2013) The role of oxidative stress in acetylcholine-induced relaxation of endothelium-denuded arteries. J Physiol Pharmacol 64:241–247

14. Cacanyiova S, Berenyiova A, Kristek F et al (2016) The adaptive role of nitric oxide and hydrogen sulphide in vasoactive responses of thoracic aorta is triggered already in young spontaneously hypertensive rats. J Physiol Pharmacol 67:501–512

15. Kubo S, Doe I, Kurokawa Y et al (2007) Direct inhibition of endothelial nitric oxide synthase by hydrogen sulfide: contribution to dual modulation of vascular tension. Toxicology 232:138–146

16. Geng B, Cui Y, Zhao J et al (2007) Hydrogen sulphide downregulates the aortic L-arginine/nitric oxide pathway in rats. Am J Physiol Regul Integr Comp Physiol 293:R1608–R1618

17. Schleifenbaum J, Kohn C, Voblova N et al (2010) Systemic peripheral artery relaxation by KCNQ channel openers and hydrogen sulfide. J Hypertens 28:1875–1882

Chapter 8

In Vivo Measurement of H$_2$S, Polysulfides, and "SSNO$^-$ Mix"-Mediated Vasoactive Responses and Evaluation of Ten Hemodynamic Parameters from Rat Arterial Pulse Waveform

Frantisek Kristek, Marian Grman, and Karol Ondrias

Abstract

The chapter describes protocols and pitfalls in in vivo studies of drug effects on anesthetized rats. It focuses on the preparation of Na$_2$S, Na$_2$S$_4$, and "SSNO$^-$ mix" solutions for rat intravenous administration, surgical preparation of jugular vein for drug administration, and preparation of carotid and tail arteries for recording of arterial pulse waveform (APW) at high resolution. It describes evaluation of ten hemodynamic parameters from APW and measurement of apparent pulse wave velocity.

Key words Rat blood pressure, Arterial pulse waveform, Carotid artery, Jugular vein, Hemodynamic parameters, H$_2$S, Na$_2$S$_4$, SSNO

1 Introduction

The experiments using invasive methods provide in great details analyses of cardiovascular responses to new vasoactive substances that could have a therapeutic potential. The experiments enable to better understand the role of individual regulatory systems in physiological processes and the mechanisms participating in development of various pathological conditions. Many of the specific patterns received from the animal experiments are transformed into noninvasive methods providing clinically beneficial information, and at the end they allow effective prevention or delay of many cardiovascular diseases.

It has been found that H$_2$S, polysulfides, and products of the NO/H$_2$S interaction are bioactive compounds [1–4]. They influence cardiovascular system including blood pressure [1, 4–10]; however their effects on particular hemodynamic parameters are unknown.

Jerzy Bełtowski (ed.), *Vascular Effects of Hydrogen Sulfide: Methods and Protocols*, Methods in Molecular Biology, vol. 2007,
https://doi.org/10.1007/978-1-4939-9528-8_8, © Springer Science+Business Media, LLC, part of Springer Nature 2019

The information obtained from arterial pulse waveform (APW) of the shape, amplitude, and duration of the waveform can provide insight into many diseases. Studies in this field have been focused on the relationships between particular APW parameters and the function of specific components of the cardiovascular system. The aim of these studies is to find unique changes in particular APW parameter(s) related to particular diseases and/or molecular mechanism of a particular pathological state(s) [11–14]. To achieve this, one needs to have APW at high resolution and a procedure to evaluate gentle changes of APW [7, 14]. In the protocols, we described preparation of Na_2S, Na_2S_4, and "SSNO⁻ mix" solutions for rat intravenous application, recording of rat APW at high resolution, and evaluation of ten hemodynamic parameters from APW.

In this chapter, the expression "H_2S" means a mixture of H_2S, HS^-, and S^{2-} in the solutions prepared by dissolving Na_2S in H_2O or buffer. The expression "SSNO⁻ mix" is used for all products, which appears after mixing of Na_2S with S-nitrosoglutathione (GSNO) in buffer solutions, when absorbance of the solution at 412 nm has maximum. The compound Na_2S_4 is a polysulfide representative.

2 Materials

1. Ultra pure deionized H_2O (0.054 microS/cm or 18.5 MΩ/cm at 25 °C; Simplicity 185 UV, Millipore).

2. Argon gas, 5.0 (Linde Gas Germany).

3. Male Wistar rats (150–400 g).

4. Anesthetics: 20 mg/ml xylazine hydrochloride (Rometar, Zentiva, Czech Republic).

5. Zoletil 100 (tiletamine hypochloride plus zolazepam hypochloride, Virbac, France). Before use, one vial should be reconstituted in 5 ml of water for injection.

6. Fiber-optic pressure transducer, FISO-LS-2FR (tip diameter 0.66 mm), FISO FPI-LS signal conditioner, FISO EVO-2 Chassis (Hugo Sachs Elektronik, Harvard Apparatus) to measure carotid artery APV.

7. Research grade blood pressure transducer (Harvard Apparatus, USA) to measure tail artery APW (*see* **Note 1**).

8. Gas-tight Hamilton syringes (250 μl) for sample application.

9. Stock solution of sodium sulfide: 100 mM Na_2S (Na_2S anhydrous, SB01, SulfoBiotics, DojinDo, Japan) in deionized H_2O bubbled with argon gas for 20 min (*see* **Note 2**). Stock solution may be frozen and stored at −80 °C for several weeks (*see* **Note 3**). Before the experiments stock solution should be thawed

and diluted with 154 mM NaCl to a final concentration using gas-tight Hamilton syringes (*see* **Note 4**).

10. Stock solution of sodium tetrasulfide: 20 mM Na$_2$S$_4$ (SB04-10, SulfoBiotics, Dojindo, Japan) in deionized H$_2$O bubbled with argon gas for 20 min (*see* **Note 5**). Aliquot the solution immediately at 50 μl to 280 full plastic vials, and put them into −80 °C freezer for several weeks (*see* **Note 6**). Just before administration, thaw one plastic vial having 20 mM Na$_2$S solution, dilute it to desired concentration with 154 mM NaCl, and use gas-tight Hamilton syringes (250 μl) for immediate administration. Discard the rest of the 20 mM Na$_2$S$_4$ solution (*see* **Note 7**, Fig. 2).

11. Stock solution of GSNO: 10 mM *S*-nitrosoglutathione (GSNO) in H$_2$O (*see* **Note 8**). Stock solution may be kept frozen at −80 °C for several weeks (*see* **Note 9**). Before the experiment, stock solution of GSNO should be thawed and diluted with 200 mM Tris–HCl, pH 7.4, to 2 mM GSNO concentration (*see* **Note 10**).

12. "SSNO$^-$ mix" solution: 1 mM GSNO, 10 mM Na$_2$S in 200 mM Tris–HCl, pH 7.4 (*see* **Note 11**).

3 Methods

3.1 Anesthesia and Rat Preparation

Carry out all procedures at room temperature unless otherwise specified. Do not use heparin in any solution.

1. Anesthetize the rat with 300 μl of the Zoletil/xylazine mixture (3:1) per 250 g of body weight intraperitoneally (*see* **Note 12**).

2. Put rat into dark box and wait 10 min.

3. After the anesthesia is fully operated (~10 min), transfer the rat to a heating plate (37 °C).

4. Shave the ventral part of the neck.

5. Before surgery fix head on the plate with a rubber band. Do not fix legs (*see* **Note 13**, Fig. 4).

6. During the anesthesia monitor blood pressure, heart rate, and reflex responses to mechanical stimuli (squeeze tail with tweezers).

7. Keep the animals under general anesthesia during the whole experiment, and kill it with an overdose of Zoletil and Rometar via the jugular vein at the end of the experiments. Rat under proper anesthesia should not have irregular BP and heart rate and should not response to mechanical stimuli.

3.2 Jugular Vein Preparation

1. On the right side over the jugular vein (cranially from the clavicle), cut the skin with a scissor.

2. Separate the jugular vein from the surrounding connective tissue using blunt forceps (*see* **Note 14**, Fig. 5).

3. Ligate the vein cranially and slit it partially with scissors (*see* **Note 15**, Fig. 6).

4. Prepare the 1-cm long of appropriate outer diameter (depending on body weight of the animal) having the rigid wall (it is important because after ligation in the vein it would not be closed). Fuse it into the soft wall 1.5-cm long cannula with higher diameter.

5. Insert the cannula into the gap and ligate it in the vein (*see* **Note 16**, Fig. 7).

6. Rinse the cannula with 100 μl of physiological saline, and clamp it with bulldog clamp until drug application.

7. Settle the needle of Hamilton syringe for drug application.

3.3 Common Carotid Artery Preparation

1. Cut the skin on the neck close to the medial line (right from the thyroid eminence).

2. Using blunt forceps pull the muscles aside, and carefully remove the left carotid artery from the connective tissue (especially be careful about nerve bundles lying close to the artery) (*see* **Note 17**, Fig. 8).

3. Strongly ligate the artery cranially.

4. Clamp the artery with bulldog clamp proximally and partially slit it between.

5. Carefully insert the fiber-optic pressure transducer, FISO-LS-2FR, into the artery, and ligate it (strong enough to stop bleeding around the catheter but allow to move the catheter in the arterial lumen). Use pressure transducer of the tip diameter 660 μm (adult rats) or transducer of the tip diameter 297 μm (for very young animals at about 60–10 g of body weight) (*see* **Note 18**, Fig. 9).

6. Immediately remove the bulldog clamp to prevent thrombus formation.

7. Put few drops of physiological saline solution (154 mM NaCl) on incision, and cover it with wet gauze.

3.4 Tail Artery Preparation

1. Cut the skin on the ventral part of the tail over the tail artery.

2. Cut the fascia that covers the artery and clean the artery of the connective tissue.

3. Ligate the artery distally and clamp with bulldog clamp proximally.

4. Slit the artery between and insert the cannula connected to the transducer to measure APW.

3.5 Detection of APW

1. Prepare fiber-optic pressure transducer, FISO-LS-2FR, FISO FPI-LS signal conditioner, FISO EVO-2 Chassis, and computer and acquisition program according to manufactory instruction. Start record signal from the fiber-optic pressure transducer, and record continuously during all experiment. Filter the recorded analog signal of APW with low-pass filter of 2.5 kHz, digitize at 10 kHz (USB-6221, National Instrument, USA), and store on a computer using the DEWESoft-7 program for data acquisition (*see* **Note 19**).

2. Insert fiber-optic pressure transducer into the artery as described above.

3. Remove the bulldog clamp and observe the recorded APW on a computer screen.

4. Wait 20–40 min until APW is stabilized, i.e., tail blood pressure is not changed during 5 min (*see* **Note 20**, Fig. 10).

5. Apply studied compounds (*see* below).

6. Export digitized data in MatLab format for further analysis.

7. At the end of the experiments, unbind the ligation, and remove the pressure transducers carefully.

8. Clean the fiber-optic pressure transducer according to manufacture instruction.

3.6 Drug Administration

1. Use 500 μl solution/kg body weight for bolus drug administration.

2. Apply all solutions from the Hamilton syringe continuously for 15 s. Using these conditions, concentration of drug in Hamilton syringe is 2000 times higher than defined dose in μmol/kg body weight. For example, if the rat has 200 g, the dose of 1 μmol/kg means that 100 μl of 2 mM solution is administered.

3. After preparation and stabilization of BP (20–40 min, Fig. 10), take the freshly prepared drug solution into gas-tight Hamilton syringe, and immediately start administrating it for 15 s via the cannulated jugular vein (*see* **Notes 20** and **21**).

4. After the integrated response is finished (the value of blood pressure returns to basal level), wash the cannula with 100 μl of physiological saline solution. After BP stabilization (≥10 min), the second and the third drug administration can be performed (*see* **Note 22**, Fig. 11).

5. Consistently control the anesthesia during the experiment.

Table 1
Definitions and abbreviations of parameters evaluated from the APW depicted from Fig. 12

Description	Abbreviation and formula
Systolic arterial BP (c or d) in (mmHg)	sBP
Diastolic arterial BP (a) in (mmHg)	dBP
Heart rate in (min^{-1})	$HR = 60/(g - a)$
Pulse blood pressure in [mmHg]	$PP = c - a$ or $d - a$
dP/dt_{max} (b) in (mmHg/ms)	dP/dt_{max}
Relative level of dP/dt_{min} (e)	$dP/dt_{min} - RL = (e - a)/(c(\text{or } d) - a)$
dP/dt_{min} (e) in (mmHg/ms)	dP/dt_{min}
Systolic area in (mmHg·s) at $a = 0$ mmHg	S-area = integral BP of a to f
Diastolic area in (mmHg·s) at $a = 0$ mmHg	D-area = integral BP of f to g
Dicrotic notch (f) relative level	$DN\text{-}RL = (f - a)/(c(\text{or } d) - a)$

3.7 Evaluation of Pulse Wave Parameters of Rat Pulse Carotid Artery Blood Pressure

1. Use DEWESoft 7.1 program to export recorded digitized data in MatLab 5.0 format for further analysis (*see* **Note 19**).

2. Use home-developed program in MatLab to find positions of seven points at APW and their time-dependent changes during experiment at 100 μs resolution. Store table of data for further analysis (*see* **Note 23**, Fig. 12).

3. Use the data to calculate ten hemodynamic parameters and their time-dependent changes (*see* **Note 24**, Table 1, Fig. 13).

3.8 Measurement of Apparent Pulse Wave Velocity

1. To study apparent pulse wave velocity, record APW from the carotid artery and tail artery. Determine the apparent pulse wave velocity as the time delay between the two pulse waves.

2. Use the table data obtained from MatLab program. Subtract time of minimum of diastolic BP from carotid artery APW (point **a**, Fig. 12) and from diastolic minimum obtained from the tail artery (Fig. 14), and calculate delay in ms of the minimum of tail artery (*see* **Note 25**) as the time delay between the two pulse waves.

3. Measure distance in mm between points of carotid and tail artery BP measurements.

4. Calculate apparent pulse wave velocity (PWV): PWV = (distance/delay) in m/s.

4 Notes

1. One can also use fiber-optic pressure transducer, FISO-LS-0.9FR (tip diameter 0.3 mm), FISO FPI-LS signal conditioner, FISO EVO-2 Chassis (Hugo Sachs Elektronik, Harvard Apparatus).

2. Na_2S can be obtained from different sources, e.g., Sigma-Aldrich.

3. Aliquot the stock solution of Na_2S immediately in 280 μl full plastic vials, and put them into −80 °C freezer for several weeks. Freshly prepared Na_2S solution can also be used.

4. The main problem working with H_2S, Na_2S_4, and "SSNO⁻ mix" is their instability in solutions under air. The best would be to perform in vivo physiological experiments without "outside oxygen" [15]. However, to keep the compounds and solutions, performing sample preparation in a glove box under argon is not very practical and might not be possible in some physiological laboratories. Therefore, in the described procedures, we used compromised protocols. The problem of impurities of NaHS and Na_2S, as a source of H_2S, was discussed by several authors [16, 17], and purity of them improved significantly as one can check it informatively by UV-Vis spectrometry [17]. When the stock Na_2S solution in H_2O is dissolved and diluted into an application buffered solution, one must be very careful, since several biologically active products (e.g., polysulfides) are formed. For example, Fig. 1 shows a

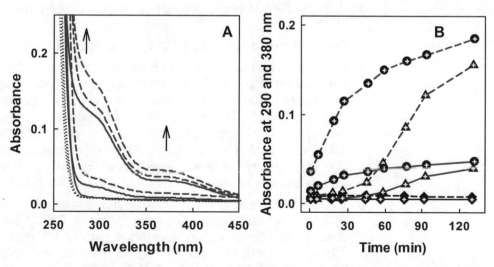

Fig. 1 (a) Increase of UV-Vis spectra of Na_2S 1 mM (black-dotted line), 2 mM (red-full line), and 5 mM (blue-dash line) after 100 mM Na_2S in H_2O was diluted in 200 mM Tris–HCl, 7.4 pH buffer, and measured at 1, 45, and 93 min at 23 °C after the dilution. **(b)** Time dependence of absorbance of Na_2S 1 mM (black diamond), 2 mM (red triangle), and 5 mM (blue circle) at 290 nm (dash line) and 380 nm (full line)

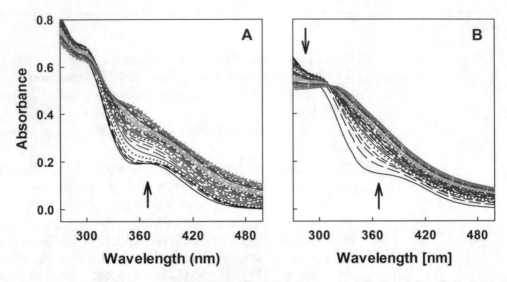

Fig. 2 Time resolved UV-Vis spectra of 200 μM Na_2S_4 in H_2O, prepared by dilution of 20 mM Na_2S_4 (in H_2O) with H_2O. The spectra were measured every 20 s for 20 min at 26 °C (**a**) and 37 °C (**b**). The increase and decrease of absorbance in time are marked by arrows

formation of polysulfides, as a time-dependent increase of absorbance at 280–400 nm ($\sim H_2S_n$) after dilution of 100 mM Na_2S in H_2O into 200 mM Tris–HCl, 7.4 pH buffer. The absorbance significantly increased at ≥ 1 min of 5 mM Na_2S and ≥ 30 min of 2 mM, but non-significantly at 1 mM stock solutions incubated at 23 °C. Time-dependent stability of application solution should be controlled by UV-Vis spectrometry.

5. Prepare 100 mM Na_2S_4. Dissolve 17.4 mg of Na_2S_4 in 1 ml of H_2O.

6. Freshly prepared Na_2S_4 solution can also be used.

7. Polysulfides in solutions under air are very unstable. For example, diluting of 20 mM Na_2S_4 solutions prepared under argon to 200 μM Na_2S_4 with H_2O under air is not stable in time as shown in Fig. 2. The diluted Na_2S_4 solution was stable for 1–2 min at 26 °C and <20 s at 37 °C. The stability of Na_2S_4 solutions in buffers is often <20 s at room temperature. Using deoxygenated buffers mostly improves stability of Na_2S_4 solution, but it should be controlled by UV-Vis spectrometry.

8. Protect solution from light. Control concentration of GSNO by UV-Vis spectrometry. In aqueous solutions, GSNO has the characteristic S-nitrosothiol absorption peak in UV region at 335 nm ($\varepsilon = 922$ M^{-1} cm^{-1}) [18]. In some studies "SSNO$^-$ mix" having different ration of H_2S/GSNO is needed. For example, to prepare 1 mM SSNO mix solution having molar ratio H_2S/GSNO = 2:1, dilute 10 mM GSNO with 200 mM Tris–HCl, pH 7.4, to 2 mM GSNO, and put 140 μl of 2 mM

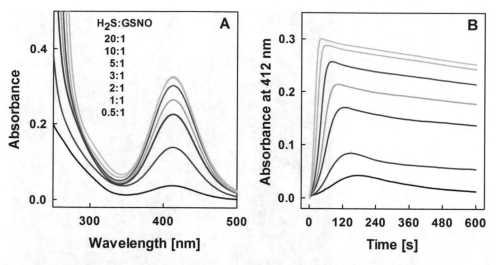

Fig. 3 (**a**) Representative UV-Vis spectra of 200 μM GSNO plus Na₂S at different molar ratios in 200 mM Tris–HCl, 7.4 pH, measured at 23 ± C. (**b**) Time-dependent absorbance at 412 nm of the samples showed at (**a**). *See* also [4, 19]

GSNO solution into 280 μl Eppendorf vial. Add 112 μl of 200 mM Tris–HCl, pH 7.4. Then add 28 μl of 20 mM Na₂S (final concentration 1 mM GSNO and 2 mM Na₂S in 280 μl vial). Close the vial and incubate at 24 °C for 2 min.

9. To store 10 mM GSNO solution for several weeks at −80 °C, use metal chelator diethylenetriaminepentaacetic acid (DTPA, 100 μM in 10 mM GSNO solution).

10. Fresh prepared GSNO solution can also be used.

11. Mixing of H₂S with NO donor GSNO produces several species, and their time-dependent concentrations depend on H₂S/GSNO ratio, buffer, pH, and temperature [4, 19–21]. In particular temperature, pH, or buffer condition, one needs to control "SSNO⁻ mix" production by UV-Vis spectra and define time of incubation, when absorbance at 412 nm has the maximum. Different H₂S/GSNO ratios might be useful in particular experiments; *see* Fig. 3 for time development of 412 nm maximum for "SSNO⁻ mix".

12. Determine the experimental volume of the administrated anesthetic solution. In case of increase and fluctuation of blood pressure and reflex responses to mechanical stimuli during experiment, reinject 100 μl of the anesthetic solution intramuscularly (i.m.). For deep anesthesia we used the mixture of 80 mg/kg Zoletil and 5 mg/kg Xylazine.

13. Rat with fixed head and free legs (Fig. 4).

14. Jugular vein separated from the environmental connective tissue (Fig. 5).

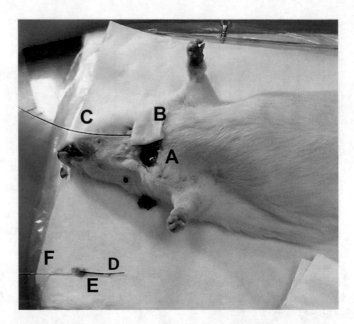

Fig. 4 Rat after experiment. (**a**) Surgery of jugular vein. (**b**) Cannulated carotic artery covered with wet gauze. (**c**) Fiber-optic cable. (**d**) Cannula having rigid wall 1 cm long. (**e**) Soft wall cannula 1.5 cm long fused with the 1 cm cannula. (**f**) Needle

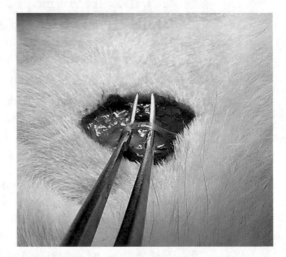

Fig. 5 Jugular vein separated from the environmental connective tissue

15. Liberated jugular vein slitting with scissors to allow insertion of the cannula (Fig. 6).

16. Prepared cannula is inserted into the slitted jugular vein (Fig. 7).

17. Left common carotid artery after liberation from the connective tissue (Fig. 8).

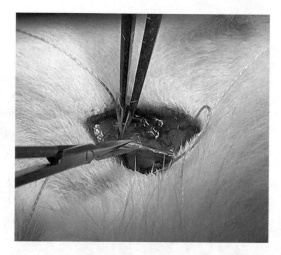

Fig. 6 Liberated jugular vein slitting with scissors to allow insertion of the cannula

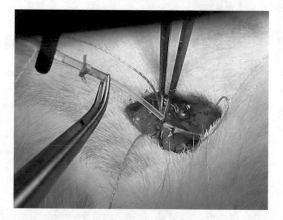

Fig. 7 Prepared cannula is inserted into the slitted jugular vein

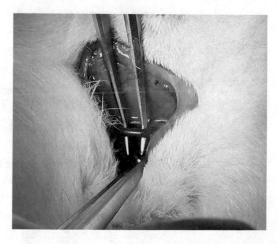

Fig. 8 Left common carotid artery after liberation from the connective tissue

Fig. 9 Pressure transducer fixed in the left common carotid artery

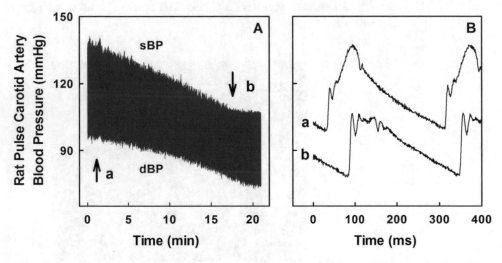

Fig. 10 The values of systolic and diastolic blood pressure (**A**) and APW (**B**) recorded from left carotid artery just after surgery (**a**) and after rat stabilization (**b**)

18. Pressure transducer fixed in the left common carotid artery (Fig. 9).

19. One can also use other devices to record, filter, digitize, and output the digitalized data in MatLab format for further analysis.

20. Stabilization of APW (Fig. 10).

21. Dead volume of cannula (i.d. 0.3 mm, 10 mm long) is 0.7 μl and it is negligible.

22. Control APW parameters if they are stabilized before repeated drug administration. Stabilization of BP does not mean that other APW parameters are stabilized as well [14] (Fig. 11).

23. Program is not yet commercially available, but other programs may recognize points on APW (Fig. 12).

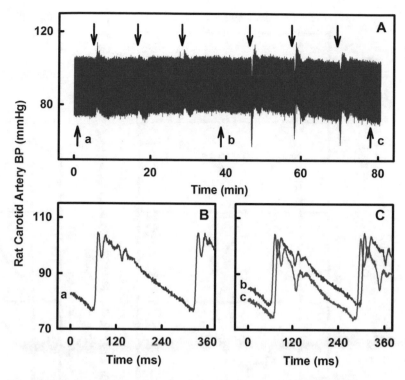

Fig. 11 (**A**) Time dependence of carotid artery BP and repetitive administration of ~1–6 μM/kg H₂S/polysulfide mixture (arrows pointing down). (**B**) APW at point "**a**." (**C**) APW at points "**b**" and "**c**." APW parameters are not stabilized at "**c**" point

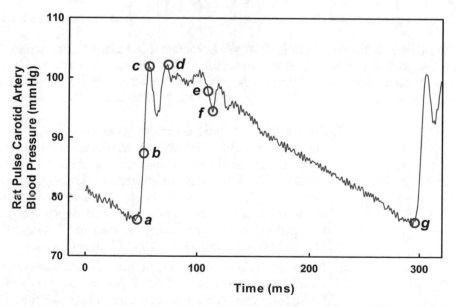

Fig. 12 The left carotid arterial pulse waveform with APW characteristics in the anesthetized rat. Diastolic blood pressure (**a, g**); dP/dt_{max} (**b**); systolic blood pressure (**c** or **d**); dP/dt_{min} (**e**); dicrotic notch (**f**)

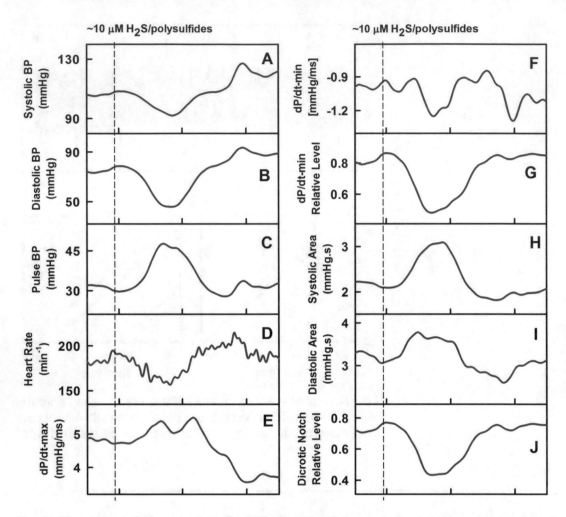

Fig. 13 Time-dependent changes of carotid APW parameters before and after i.v. bolus administration of ~10 µM H_2S/polysulfide mixture obtained after incubation of H_2S in buffer as it is indicated by a dashed vertical line. For abbreviations and description of the parameters, *see* Table 1 and Fig. 12. Stock solution of 20 mM Na_2S was incubated for 7 min at 24 °C in 200 mM Tris–HCl, 7.4 pH buffer before bolus administration

24. From the known time-dependent position of 7 points of APW, one can calculate numerous hemodynamic parameters and their time development under the influence of drugs (Table 1, Fig. 13). In our recent studies, 12 and 14 parameters were used [7, 14]. Table 1 shows an example of definition of ten hemodynamic parameters and time-dependent effect of H_2S-polysulfide mixture obtained after incubation of 10 µM H_2S in 200 mM Tris–HCl, 7.4 pH buffer (Fig. 13).

25. Evaluation of time-dependent apparent pulse wave velocity and correlating it with carotid APW parameters may find a new biological connection between them (Figs. 13 and 14). For the connection of apparent pulse wave velocity to arterial stiffness, *see* [4, 7].

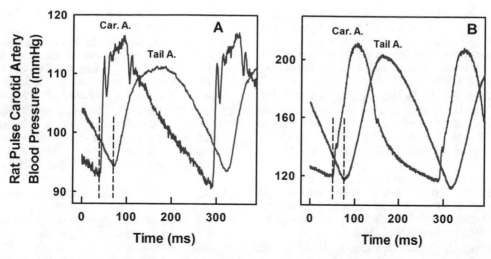

Fig. 14 The pulse wave recorded from the left carotid artery (red) and from tail artery (blue) before (**a**) and after administration 10 µg/kg phenylephrine (**b**). The time delay is marked by dashed vertical lines. Blood pressure values are valid for carotid artery only. To see delay of the pulse wave from tail artery, it was artificially normalized

Acknowledgments

This work was supported by the Slovak Research and Development Agency under contract No. APVV-15-0371 and financial support by Slovak grants VEGA 2/0067/13, 2/0048/17, 2/0079/19, Ministry of Health of the Slovak Republic under the project registration number 2012/51-SAV-1 and APVV-15-0565.

References

1. Wang R (2012) Physiological implications of hydrogen sulfide: a whiff exploration that blossomed. Physiol Rev 92:791–896

2. Filipovic MR, Eberhardt M, Prokopovic V, Mijuskovic A, Orescanin-Dusic Z, Reeh P et al (2013) Beyond H$_2$S and NO interplay: hydrogen sulfide and nitroprusside react directly to give nitroxyl (HNO). A new pharmacological source of HNO. J Med Chem 56:1499–1508. https://doi.org/10.1021/jm3012036

3. Kimura H (2014) Hydrogen sulfide and polysulfides as biological mediators. Molecules 19:16146–16157. https://doi.org/10.3390/molecules191016146

4. Cortese-Krott MM, Kuhnle GGC, Dyson A, Fernandez BO, Grman M, DuMond JF et al (2015) Key bioactive reaction products of the NO/H2S interaction are S/N-hybrid species, polysulfides, and nitroxyl. Proc Natl Acad Sci U

S A 112:E4651–E4660. https://doi.org/10.1152/physrev.00017.2011

5. Zhao W, Zhang J, Lu Y, Wang R (2001) The vasorelaxant effect of H2S as a novel endogenous gaseous KATP channel opener. EMBO J 20:6008–6016

6. Zhang Z, Huang H, Liu P, Tang C, Wang J (2007) Hydrogen sulfide contributes to cardioprotection during ischemia-reperfusion injury by opening KATP channels. Can J Physiol Pharmacol 85:1248–1253

7. Tomasova L, Pavlovicova M, Malekova L, Misak A, Kristek F, Grman M et al (2015) Effects of AP39, a novel triphenylphosphonium derivatised anethole dithiolethione hydrogen sulfide donor, on rat haemodynamic parameters and chloride and calcium Cav3 and RYR2 channels. Nitric Oxide 46:131–144. https://doi.org/10.1016/j.niox.2014.12.012

8. Drobna M, Misak A, Holland T, Kristek F, Grman M, Tomasova L (2015) Captopril partially decreases the effect of H_2S on rat blood pressure and inhibits H_2S-induced nitric oxide release from S-nitrosoglutathione. Physiol Res 64:479–486

9. Tomasova L, Dobrowolski L, Jurkowska H, Wróbel M, Huc T, Ondrias K et al (2016a) Intracolonic hydrogen sulfide lowers blood pressure in rats. Nitric Oxide 60:50–58. https://doi.org/10.1016/j.niox.2016.09.007

10. Tomasova L, Konopelski P, Ufnal M (2016b) Gut bacteria and hydrogen sulfide: the new old players in circulatory system homeostasis. Molecules 21:pii: E1558. https://doi.org/10.3390/molecules21111558

11. Avolio AP, Butlin M, Walsh A (2010) Arterial blood pressure measurement and pulse wave analysis—their role in enhancing cardiovascular assessment. Physiol Meas 31:R1–R47

12. Stoner L, Young JM, Fryer S (2012) Assessments of arterial stiffness and endothelial function using pulse wave analysis. Int J Vasc Med 2012:903107

13. Zhang J, Critchley LA, Huang L (2015) Five algorithms that calculate cardiac output from the arterial waveform: a comparison with Doppler ultrasound. Br J Anaesth 115:392–402. https://doi.org/10.1093/bja/aev254

14. Tomasova L, Kristek F, Grman M, Ondriasova E., Ondrias K (2015) Effects of the reaction products of sulfide and Snitrosoglutathione on rat hemodynamic parameters. Nitric Oxide 47: S30–S31. https://doi.org/10.1016/j.niox.2015.02.074

15. Wedmann R, Bertlein S, Macinkovic I, Böltz S, Miljkovic JL, Muńoz LE et al (2014) Working with "H2S": facts and apparent artifacts. Nitric

Oxide 41:85–96. https://doi.org/10.1016/j.niox.2014.06.003

16. Hughes MN, Centelles MN, Moore KP (2009) Making and working with hydrogen sulfide: the chemistry and generation of hydrogen sulfide in vitro and its measurement in vivo: a review. Free Radical Biol Med 47:1346–1353

17. Nagy P, Pálinkás Z, Nagy A, Budai B, Tóth I, Vasas A (2014) Chemical aspects of hydrogen sulfide measurements in physiological samples. Biochim Biophys Acta 1840:876–891. https://doi.org/10.1016/j.bbagen.2013.05.037

18. Hart TW (1985) Some observations concerning the S-nitroso and S-phenylsulphonyl derivatives of L-cysteine and glutathione. Tetrahedron Lett 26:2013–2026

19. Grman M, Misak A, Jacob C, Tomaskova Z, Bertova A, Burkholz T et al (2013) Low molecular thiols, pH and O_2 modulate H_2S-induced S-nitrosoglutathione decomposition—•NO release. Gen Physiol Biophys 32:429–441. https://doi.org/10.4149/gpb_2013026

20. Filipovic MR, Miljkovic JL, Nauser T, Royzen M, Klos K, Shubina T, Koppenol WH, Lippard SJ, Ivanović-Burmazović I (2012) Chemical characterization of the smallest S-nitrosothiol, HSNO; cellular cross-talk of H_2S and S-nitrosothiols. J Am Chem Soc 134:12016–12027. https://doi.org/10.1021/ja3009693

21. Cortese-Krott MM, Fernandez BO, Santos JL, Mergia E, Grman M, Nagy P et al (2014) Nitrosopersulfide (SSNO(-)) accounts for sustained NO bioactivity of S-nitrosothiols following reaction with sulfide. Redox Biol 2:234–244. https://doi.org/10.1016/j.redox.2013.12.031

Chapter 9

Simultaneous Measurements of Tension and Free H₂S in Mesenteric Arteries

Elise Røge Nielsen, Anna K. Winther, and Ulf Simonsen

Abstract

Hydrogen sulfide (H_2S), in addition to nitric oxide and carbon monoxide, is the third gasotransmitter and known to cause relaxation in peripheral arteries. Here we describe a method that allows simultaneous measurement of contractility in arteries mounted in an isometric wire myograph and the concentration of free H_2S in the lumen of the artery as well as in the organ bath. This method can be used to directly correlate how much free H_2S is needed to cause relaxation, which previously has been difficult to answer as H_2S can be found in many different forms.

Key words Hydrogen sulfide, Tension, Resistance arteries, Myography, Sensor

1   Introduction

The regulation of peripheral resistance is to a large degree determined by the tone of small arteries, such as the mesenteric small arteries. It is possible to investigate the function of the arteries in vitro after dissection and mounting of them in a wire myograph, which allows pharmacological examination of the mechanisms involved in, for instance, relaxation to hydrogen sulfide (H_2S). H_2S is a gasotransmitter known to cause relaxation of arteries, and therefore novel H_2S donors are suggested to be potential new antihypertensive drugs [1].

The precise role of H_2S in the regulation of vessel contractility is still somewhat unclear. In rat arteries it has been suggested that low H_2S concentrations cause contractions, while higher H_2S concentrations have been shown to induce relaxation of a number of both rat and human arteries, among these the rat mesenteric arteries [2–5]. The physiological role of H_2S in the control of arterial tone has been debated, since the concentration of added Na_2S or NaHS needed to cause relaxation in the arteries is very high compared to physiological levels. Sulfide can be distributed as a variety of species, H_2S, HS^-, and S^{2-}, determined by pH

Jerzy Bełtowski (ed.), *Vascular Effects of Hydrogen Sulfide: Methods and Protocols*, Methods in Molecular Biology, vol. 2007, https://doi.org/10.1007/978-1-4939-9528-8_9, © Springer Science+Business Media, LLC, part of Springer Nature 2019

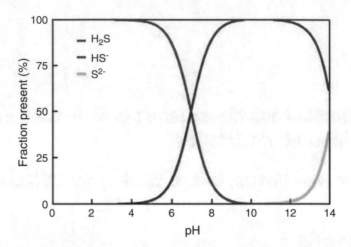

Fig. 1 The distribution of H_2S, HS^-, and S^{2-} as a function of pH

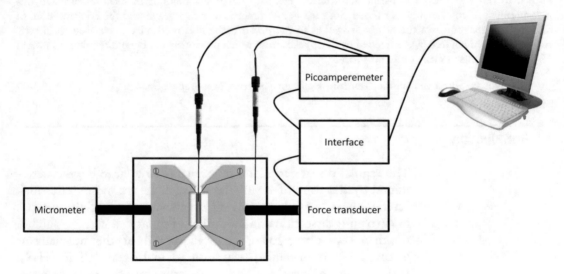

Fig. 2 Schematic drawing of the setup

(*see* Fig. 1). Sulfides are also reported to interact with radical oxygen species and to bind to transition metal ions [6]. The methods which have been used to determine sulfide amounts are either not specific to H_2S or not possible to use in measuring the amount of H_2S in the lumen of the artery (*see* Nagy [6] for a review of other sulfide measurements). Therefore a method is needed which couples the relaxation directly to the amount of free H_2S to investigate if the physiological levels of H_2S have any relevance to arterial tone (Fig. 2).

Direct measurement of [H_2S] is possible with a H_2S-sensitive microsensor, which is a miniaturized amperometric sensor consisting of an internal reference and a sensing and guard anode. H_2S from the environment is driven by the external partial pressure and

Fig. 3 H$_2$S sensor from Unisense, Aarhus, Denmark

will penetrate through the sensor tip membrane into the alkaline electrolyte, and the HS$^-$ ions formed are oxidized immediately by ferricyanide, producing sulfur and ferrocyanide. The latter is electrochemically reoxidized at the exposed end of the platinum working electrode, thereby creating a current that is directly proportional to the dissolved H$_2$S concentration at the sensor tip (Fig. 3) [7]. We have published results from simultaneous measurements of tension and free H$_2$S showing that addition of 300 μM Na$_2$S resulted in 15 μM of free H$_2$S in the lumen of the artery and 6.5 μM in the organ bath, which was sufficient to cause 60% relaxation of mesenteric arteries. It appears as the concentration is higher in the lumen of the artery than in the organ bath showing the relevance of measuring at both places [8]. In this chapter, we will describe this method, which combines the examination of isolated arteries in an isometric wire myograph with simultaneous measurements of H$_2$S in the organ bath and in the lumen of the artery. We will describe the isolation of the superior mesenteric artery, the mounting procedure, construction of the concentration response curves, calibration, and insertion of the sensor as well as data handling.

2 Materials

1. Regular microscope.

2. Microforceps, microscissors, and petri dish for dissection.

3. Single channel wire myograph (e.g., model 320A, Danish Myo Technology, Aarhus, Denmark) adjusted with a hole to insert the electrode [9].

4. Computer and software (e.g., LabScribe, World Precision Instruments, Hitchin, UK).

5. H$_2$S-sensitive microelectrodes with tip diameters of 50–80 μm (Unisense A/S, Aarhus, Denmark).

6. H$_2$S-sensitive microelectrode with needle tip (Unisense A/S, Aarhus, Denmark) for organ bath measurements.

7. Micromanipulator.

8. Adjustable electrode holder.

9. Picoampere meter.

10. Electric thermometer.

11. pH meter.

12. Air mixture of 20% O$_2$ and 5% CO$_2$.

13. Nitrogen gas.

14. Physiological saline solution with 1.6 mM calcium ($PSS_{1.6}$): 119 mM NaCl, 25 mM $NaHCO_3$, 5.5 mM glucose, 1.6 mM $CaCl_2$, 1.18 mM KH_2PO_4, 1.17 mM $MgSO_4$, 0.027 mM ethylenediaminetetraacetic acid (EDTA).

15. Physiological saline solution 0.0 mM calcium ($PSS_{0.0}$), composition as PSS 1.6 but without $CaCl_2$.

16. Noradrenaline (NA).

17. Acetylcholine (Ach).

18. Hydrochloride acid 1 M.

19. 1 M Na_2S solution (made fresh every day and the solution is made in a fume hood) in $PSS_{0.0}$ pre-bubbled with N_2 (*see* **Note 1**), pH adjusted to 7.35–7.45 by adding 1 M hydrochloric acid (*see* **Note 2**). Keep solution in a closed Nunc tube on ice until use.

20. Ethanol 90%.

21. Acetic acid 8%.

22. MilliQ water.

3   Methods

3.1   Calibration of the Electrodes

1. Connect the H_2S sensor to a Unisense picoampere amplifier, and allow the sensor a period of polarization before use (*see* **Note 3**).

2. Connect the picoampere meter to the interface (*see* **Note 4**).

3. Preheat five tubes with 40 mL N_2 bubbled PSS adjusted to pH below 4 with hydrochloric acid to 37 °C (*see* **Notes 5–7**).

4. Dilute the Na_2S solution to 10 mM with PSS.

5. Add 40 μl of 10 mM Na_2S solution to the 40 ml PSS to reach a final concentration of 10 μM.

6. Place the sensors in the solution and allow the signals to stabilize.

7. In the next tube, add 80 μl of 10 mM Na_2S solution to the 40 ml PSS to reach a final concentration of 20 μM.

8. Place the sensors in the solution and allow the signal to stabilize.

9. To the next tube, add 120 μl of 10 mM Na_2S solution to the 40 ml PSS to reach a final concentration of 30 μM.

10. Place the sensors in the solution and allow the signal to stabilize.

11. To the next tube, add 200 μl of 10 mM Na_2S solution to the 40 ml PSS to reach a final concentration of 50 μM.

12. Place the sensors in the solution and allow the signal to stabilize.

13. Rinse the sensors in milliQ water.

3.2 Dissection of the Superior Mesenteric Artery

1. Sacrifice the rats by cervical dislocation followed by exsanguination.

2. Remove the intestine including the superior mesenteric artery and the kidneys.

3. Place the mesenteric bed into a petri dish with a thick layer of clear silicone gel in base, and cover with ice-cold PSS.

4. Pin each of the kidneys and identify the superior mesenteric artery between the kidneys.

5. Carefully remove connective tissue, and dissect the artery using a microscope, microforceps, and microscissors (*see* **Note 8**).

6. Cut the artery into segments of approximately 2 mm, where one end is cut obliquely (*see* **Notes 9** and **10** and Fig. 4).

3.3 Mounting of the Mesenteric Artery (Fig. 5)

1. Close the two jaws around the first 100 μm stainless steel wire to hold it in place, while it is fixed to the left jaw using the top screw (*see* **Note 11**).

2. Put the artery segment on the wire and open the jaws to place the segment in between the jaws.

Fig. 4 Schematic of the vessel segment

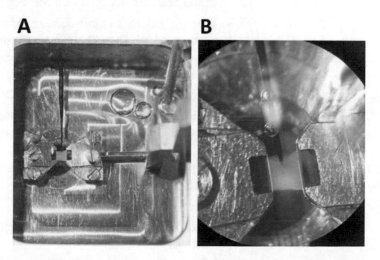

Fig. 5 (**a**) Mesenteric artery mounted on the myograph with a sensor placed in the lumen. (**b**) Close-up of the artery with sensor inserted in the lumen

3. When the artery is in place, gently close the jaws again, and fix the wire to the bottom screw on the left jaw (*see* **Note 12**).

4. Open the jaws and tighten the screws again.

5. Gently put the second wire through the lumen of the segment using microforceps (*see* **Note 13**).

6. Close the jaws again while making sure that the second wire goes below the first to avoid them crossing over.

7. Fix the second wire to the right jaw using the two screws (*see* **Note 12**).

8. Slightly open the jaws and adjust the wire so they are in the same horizontal plane.

9. Adjust jaws so the wires are almost touching.

10. Measure the length of the segment using a calibrated eyepiece in the dissecting microscope. Place one dotted line at each end of the vessel segment and measure (*see* **Note 14**).

11. Let the segment equilibrate for 30 min at 37 °C while bubbled with 5% CO_2/21% O_2/74% N_2.

3.4 Normalization of the Superior Mesenteric Artery

1. Turn on LabScribe software (LabScribe2, World Precision Instruments, Sarasota, FL, USA) and start recording.

2. Note the initial micrometer reading (X_0) and the corresponding force reading (Y_0) when the wires are almost touching in Excel. This will correspond to zero wall force.

3. Slowly move wires apart using the micrometer screw, and after 1 min note a new micrometer (Y) and force (X) reading in the Excel to calculate the corresponding transmural pressure as described by Mulvany and Halpern [10].

4. Repeat this procedure approximately five times until the transmural pressure exceeds 100 mmHg (13.3 kPa).

5. Plot the internal circumference of the vessel against the wall tension, which should give an exponential curve. Calculate the point at which the curve intersects the 100 mmHg isobar as the internal circumference (IC_{100}).

6. Calculate IC_1 as 90% of IC_{100} (*see* **Note 15**).

7. Adjust the micrometer to the IC_1 value.

8. Let the segment to equilibrate for 30 min.

3.5 Viability Test of the Superior Mesenteric Artery

1. Add 10 μl of 1 mM NA to the organ bath (final concentration 1 μM).

2. Wait for the signal to stabilize.

3. Add 5 μl of 1 mM M NA to the organ bath (final concentration 500 nM).

4. Wait for the signal to stabilize.

5. Add 10 μl 1 mM acetylcholine (Ach) to the organ bath (final concentration 1 μM) (*see* **Note 16**).

6. Wash with PSS$_{1.6}$ (*see* **Note 17**).

3.6 Insertion of the Electrodes

1. Carefully insert the microelectrode (50–80 μm tip) into the micromanipulator (*see* **Note 18**).

2. Remove the cap covering the hole into the organ bath in the myograph.

3. Carefully move the microelectrode through the opening into the organ bath using the micromanipulator.

4. Use the microscope and micromanipulator to place the microelectrode 1 mm inside the lumen of the artery.

5. The needle tip sensor is placed in an adjustable electrode holder and placed in the organ bath (*see* **Note 19**).

6. Adjust the bubbling of the organ bath (*see* **Note 20**).

7. Replace the cap with grease around the micromanipulator.

3.7 Construction of the Concentration Response Curve

1. Contract the vessel segment with 10 μl 1 mM M NA (final concentration 1 μM), and wait till contraction is stable.

2. Add 10 μl of 1 mM Na$_2$S to the organ bath (final concentration 1 μM), and wait till contraction level is stable.

3. Continue adding Na$_2$S to the organ bath cumulatively with final concentrations of 3 μM, 5 μM, 10 μM, 30 μM, 50 μM, 100 μM, 300 μM, 500 μM, 1 mM, 3 mM, and 5 mM (*see* **Notes 21** and **22**).

3.8 End of Experiment and Cleaning

1. Remove the H$_2$S sensor from the lumen and organ bath using the micromanipulator.

2. Rinse the sensor with demineralized water and place them in a secure place (*see* **Notes 23** and **24**).

3. Remove the artery from the myograph.

4. Clean the myograph by filling the chamber with 90% EtOH and bubble for a few minutes.

5. Clean organ bath with a cotton stick while 90% EtOH is in the organ bath.

6. Remove EtOH.

7. Fill the organ bath with 8% acetic acid.

8. Clean the organ bath with a cotton stick while 8% acetic acid is in the organ bath.

9. Rinse the organ bath several times with demineralized water.

10. Dry the organ bath using small pieces of paper towel and microforceps.

11. Replace old grease with new grease.

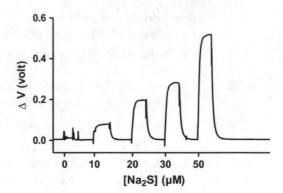

Fig. 6 Trace of sensor calibration

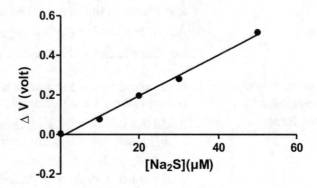

Fig. 7 Linear regression of calibration

3.9 Data Analysis

1. Extract all data from the LabScribe program.

2. Extract the signal from each of the calibration points for both sensors.

3. Make a standard curve for each H_2S sensor. Register the concentration of H_2S on the *x*-axis and the signal in voltage on the *y*-axis (*see* Fig. 6).

4. Make a linear regression to obtain an equation to convert the sensor signal in volt to $[H_2S]$ (*see* Fig. 7).

5. Calculate the signal from the sensors corresponding to the concentration response curve (Fig. 8).

6. Convert the tension of the artery (mN) to percent relaxation by setting the stable contraction to NA – baseline tension to 100%. Then convert all other points (here called *x*) to percent relaxation by using the equation:

Percent relaxation $= 100\% \times (x\,\text{mN} - \text{baseline})/(\text{Stable contraction} - \text{baseline})$.

7. The relaxation and $[H_2S]$ can now be directly related.

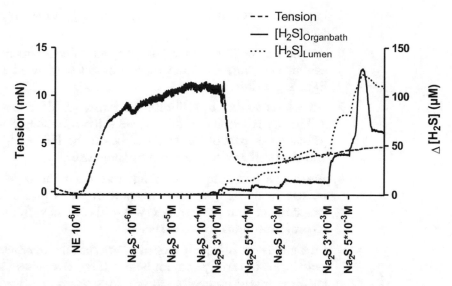

Fig. 8 Trace of original experiment showing tension, [H₂S] in lumen and [H₂S] in the organ bath

4 Notes

1. The PSS$_{0.0}$ is bubbled with N_2 to avoid interactions between Na$_2$S and O_2 in the PSS$_{0.0}$.

2. As the solution of Na$_2$S in PSS$_{0.0}$ is very basic, it can affect the tension of the artery if this is added to the myograph; therefore the solution is neutralized before use.

3. When the sensor is not in use, H_2S will build up inside the electrolyte. This must be removed by the sensing anode and the guard anode before stable operation of the sensor is possible. A period of polarization is needed to remove the buildup of H_2S from the anode and guard anode. If the sensor has not been in use for several days, allow at least 2 h of pre-polarization. If the sensor has been in use earlier, wait until the signal is stable for at least 10 min.

4. The picoampere meter can be used on its own, but when connecting it to the interface, the results are recorded in Lab-Scribe as one channel for each sensor, while the tension is recorded on a third channel. This allows direct simultaneous recording of the signals.

5. Because the sensor is sensitive to temperature, we performed all calibrations and measurements at 37 °C, as this temperature is physiologically appropriate. We use a polystyrene box filled with water heated to 37 °C to be able to maintain the calibration solutions at the right temperature without having too much electrical noise from a warm bath. Temperature was

kept at 37 °C by adjusting with boiling water from time to time.

6. As the sensor is sensitive to light, all tubes containing the calibration solutions were wrapped in tinfoil to avoid as much light as possible.

7. Always remember to calibrate the sensor at pH values below 4.0 because then all the added Na_2S will be on the H_2S form; at pH above 4, part of the H_2S will be on the HS^- or S^{2-} form and thus will not be measured by the sensor.

8. Avoid the first couple of millimeters from the aorta as the muscle tissue here is different, and as far as possible, avoid branches on your piece of artery. Avoid contact with the artery to prevent trauma to the tissue.

9. One end is cut obliquely to make sure that the microelectrode goes inside the artery and not below. If the artery has a straight cut, it is almost impossible to see if the sensor is placed in the lumen or below the artery.

10. The arterial rings should be cut into 1.5–2.0-mm-long segments with internal diameters of approximately 1000–1300 μm.

11. The wires should have a diameter of 100 μm to ensure room enough for the sensor.

12. Do not tighten the screws all the way when the jaws are still closed together, but wait till the jaws are apart to make sure it is properly tightened all the way and to avoid the wire breaking.

13. The second wire should as far as possible be inserted in the artery in a fluid motion to avoid damaging the endothelium or the smooth muscle cells in the artery segment.

14. The measurement is registered in the LabChart software together with the conversion factor from the calibration of the eyepiece to calculate the length of the segment.

15. The micrometer reading corresponding to the IC_1 value is also calculated, and the micrometer is adjusted accordingly to achieve optimal maximum active tension.

16. Relaxation to Ach should be at least 70% of NA contraction reflecting intact endothelial function.

17. If the contraction or endothelial function is not considered optimal, remove the artery and start over with a new segment to obtain optimal results.

Another description of mounting in a wire myograph can be found in [11] and the original article about the technique in [10].

18. The right placement of the micromanipulator behind the myograph is determined by using a stick instead of the microelectrode to avoid breakage of the glass tip on the microelectrode.

19. To measure [H₂S] in the organ bath, a needle tip sensor is used. The needle tip sensor can be placed in the organ bath without worrying about hitting the bottom of the bath, which would break the glass tip of the H₂S microsensor.

20. In the presence of the microelectrode, the bubbling should be set to an absolute minimum to avoid noise disturbing the sensor signal.

21. Wait 3–5 min between each addition or for stabile signal of both vessel and sensors. In our preparations, we see most of the relaxation occurring at 3×10^{-4} M Na₂S; therefore extra time for both tension and sensor signals is appropriate at this addition; thus wait for approximately 10 min at this point.

22. Another curve can be run after the first concentration response curve. If you are running a new curve, wash by exchanging the PSS in the organ bath at least three times, and wait for 30 min. After 30 min of washout time and when signals from both sensors and the force transducer are stable, another curve can be performed. Be aware that any drugs used are reversible, so they won't interact with the second curve performed.

23. If the sensors are to be used the following day, leave them connected to the picoampere meter. If the sensors are not to be used for a longer time period, disconnect and place them in a secure place.

24. Changes in the coating of the electrode cause abrupt changes in basal electrode current, and in such case the electrode has to be discarded.

References

1. van Goor H et al (2016) Hydrogen sulfide in hypertension. Curr Opin Nephrol Hypertens 25(2):107–113

2. Ali MY et al (2006) Regulation of vascular nitric oxide in vitro and in vivo; a new role for endogenous hydrogen sulphide? Br J Pharmacol 149(6):625–634

3. Kubo S et al (2007) Direct inhibition of endothelial nitric oxide synthase by hydrogen sulfide: contribution to dual modulation of vascular tension. Toxicology 232 (1-2):138–146

4. Webb GD et al (2008) Contractile and vasorelaxant effects of hydrogen sulfide and its biosynthesis in the human internal mammary artery. J Pharmacol Exp Ther 324(2):876–882

5. d'Emmanuele di Villa Bianca R et al (2011) Hydrogen sulfide-induced dual vascular effect involves arachidonic acid cascade in rat mesenteric arterial bed. J Pharmacol Exp Ther 337 (1):59–64

6. Nagy P et al (2014) Chemical aspects of hydrogen sulfide measurements in physiological samples. Biochim Biophys Acta Gen Subj 1840 (2):876–891

7. Jeroschewski P, Steuckart C, Kühl M (1996) An amperometric microsensor for the determination of H2S in aquatic environments. Anal Chem 68:7

8. Hedegaard ER et al (2016) Involvement of potassium channels and calcium-independent mechanisms in hydrogen sulfide-induced

relaxation of rat mesenteric small arteries. J Pharmacol Exp Ther 356(1):53–63

9. Simonsen U et al (1999) In vitro simultaneous measurements of relaxation and nitric oxide concentration in rat superior mesenteric artery. J Physiol 516(Pt 1):271–282

10. Mulvany MJ, Halpern W (1977) Contractile properties of small arterial resistance vessels in spontaneously hypertensive and normotensive rats. Circ Res 41(1):19–26

11. Spiers A, Padmanabhan N (2005) A guide to wire myography. In: Fennell JP, Baker AH (eds) Hypertension: methods and protocols. Humana Press, Totowa, NJ, pp 91–104

Chapter 10

The Relaxant Mechanisms of Hydrogen Sulfide in Corpus Cavernosum

Fatma Aydinoglu and Nuran Ogulener

Abstract

In several animal and human studies, the contribution of the endothelium, nitric oxide/soluble guanosine monophosphate (NO/cGMP) pathway, adenylyl cyclase, phosphodiesterase (PDE), potassium (K^+) channels, L-type calcium channels, Na^+-K^+-ATPase, muscarinic acetylcholine receptors, RhoA/Rho-kinase pathway, and cyclooxygenase (COX)/arachidonic acid cascade on the relaxant mechanism of L-cysteine/H_2S pathway in corpus cavernosum has been investigated. In this chapter the relaxant mechanisms of H_2S in corpus cavernosum is discussed with data available in the current relevant literature. Also, in vitro experimental procedure for mice corpus cavernosum which used to investigate the relaxant effect of H_2S is given in detail.

Key words Corpus cavernosum, Erectile function, Erectile dysfunction

1 Introduction

Hydrogen sulfide (H_2S) is another gasotransmitter like nitric oxide (NO) and carbon monoxide (CO) [1]. H_2S produces relaxations in vascular and other smooth muscles [2–9]. The first pilot study in primates and rats revealed the possible role of H_2S in erectile function [10]. The following studies reported the presence of L-cysteine/H_2S in erectile tissue and the contribution of this pathway in erectile function [11–16]. Also, it has been suggested that the mechanisms of H_2S regulating penile erection may be a new therapeutic approach for the treatment of erectile dysfunction [17]. In this chapter, the regulation of cavernosal tissue tone in penile erection, the role of H_2S in erectile function, and the relaxant mechanisms of H_2S in corpus cavernosum are presented. Also, in vitro experimental procedure to study the relaxant effect of H_2S in mice corpus cavernosum is explained in detail.

1.1 The Regulation of Cavernosal Tissue Tone in Penile Erection

The penile erectile apparatus consists of two vascularized paired dorsally placed corpora cavernosa, which contain the trabecular smooth muscle that line the sinusoids, and the ventral corpus

Jerzy Bełtowski (ed.), *Vascular Effects of Hydrogen Sulfide: Methods and Protocols*, Methods in Molecular Biology, vol. 2007,
https://doi.org/10.1007/978-1-4939-9528-8_10, © Springer Science+Business Media, LLC, part of Springer Nature 2019

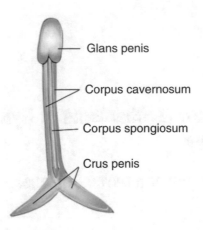

Fig. 1 The anatomical structure of mouse penis

spongiosum that surrounds the urethra and terminates as the glans
penis (Fig. 1). The cavernosal tissues are surrounded by tunica
albuginea and separated by a perforated septum [18, 19]. The
penile erectile tissue is composed of corpora cavernosa which
consist of the extensive endothelium-lined sinusoid spaces
[20]. Penile erection is a complex neurovascular process of the
corpus cavernosum smooth muscle [21], and relaxation of corporal smooth muscle is essential for normal erectile function
[22]. The relaxation of cavernosal and arterial smooth muscle
decreases vascular resistance and permits blood flow into the
sinusoidal spaces of cavernosal tissue, which physically reduces
venous outflow and increases intracavernosal pressure, resulting
in penile erection [18–20, 23, 24]. The blood flow to the trabecular spaces by cavernosal arteries is required in order to enable
penile erection. The main arterial supply to penis is pudendal
artery which splits into bulbourethral, dorsal, and cavernosal
arteries. The helicine arteries are branches from cavernosal arteries
and provide blood flow to sinusoidal spaces of corpus cavernosum
and act as resistance arteries [19, 25]. In addition, the restriction
of venous outflow from the penis is involved in maintaining erectile function [19, 20, 26]. During penile erection, enlarged
corpora cavernosa restrain emissary veins, and the capacity of
emissary veins to drain blood from the penis decreases [27]. The
nerves and endothelium of sinusoids and vessels in the penis
produce and release transmitters and modulators which interact
in their control of the contractile state of the penile smooth
muscles: detumescence and flaccidity, tumescence, and erection
[22, 28]. With regard to the mechanisms of penile erection, the
cavernous trabecular smooth muscle is considered an extension of
the systemic vasculature [29]. The impairment of the relaxant
mechanism of this smooth muscle causes impotence [30].

1.2 Hydrogen Sulfide H_2S was known as a toxic gas which though is solely produced by commensal bacteria. However, it has been determined that H_2S is synthesized enzymatically in various animal and human tissues, including vascular, heart, liver, kidney, lung, gastrointestinal, central system, reproductive organ, skeletal muscle, and adipose tissue, and endogenous H_2S may play a role as neuromodulator in the brain [1, 2, 31–41]. In recent years, H_2S has been accepted as a gas neurotransmitter in mammals like nitric oxide (NO) and carbon monoxide (CO) [1]. In common with other gasotransmitters, H_2S is a small molecule gas, endogenously synthesized through enzymes in mammalian tissues, which passes through the cell membranes without binding to a specific receptor or carrier with short half-life after releases [42, 43]. H_2S is produced from L-cysteine endogenously by three different enzymes, cystathionine gamma lyase (CSE), cystathionine beta-synthase (CBS), or 3-mercaptopyruvate sulfurtransferase (3-MST) along with cysteine amino transferase in various tissues [31, 44, 45]. The synthesis of H_2S from L-cysteine is catalyzed particularly by pyridoxal-5-'-phosphate-dependent enzymes CSE and CBS [46]. CBS and CSE enzymes are tissue-type specific; CBS is expressed in the airway tissues [8], bladder [7], brain [41], colon [34], ileum [2], kidney [47], liver [47, 48], pancreatic islets [49], placenta, uterus [38], and urothelium [50], while CSE is found in the airway tissues [8], adipose tissue [39], bladder [7], brain [51], ileum [2], penile corpus cavernosal tissue [12, 13], vascular tissues [3, 4], kidney [47, 48], liver [3, 47, 51], pancreatic islets [49], and placenta and uterus [38]. Although both enzymes are present in the brain [47] as well as in some peripheral tissues including pulmonary and airway tissues [8], CBS is the major enzyme that produces H_2S in the brain, whereas CSE is more abundant in the cardiovascular system and peripheral tissues [32, 44, 47]. In addition, 3-MST is found in the brain, as well as the heart, lung, liver, kidney, thymus, testis, and endothelium and smooth muscle of aorta [52–54]. Recently, H_2S formation from D-cysteine is reported through the 3-MST pathway, via D-amino acid oxidase which provides 3-mercaptopyruvate for 3-MST [55]. This pathway is mainly present in the cerebellum and kidney [56]. The levels of H_2S in the brain are determined to range between 50 and 160 μM [41]. Zhao and colleagues reported that plasma level of H_2S is approximately 50 μM [3]. H_2S regulates many functions in central nervous system, cardiovascular, respiratory, gastrointestinal, endocrine, and urogenital systems [57].

1.3 Hydrogen Sulfide and Corpus Cavernosum   The role of H_2S in erectile function was firstly demonstrated by Srilatha et al. [10] who have shown that intracavernosal injection of NaHS as exogenous H_2S substrate caused increases in intracavernosal pressure and penile length in primates, and the pharmacological inhibition of CSE enzyme in rat resulted in marked alleviation of cavernous nerve stimulation-induced perfusion pressure, suggesting the possible role for endogenous H_2S in erectile function through facilitation of nerve-mediated penile tumescence [10]. Also, endogenous H_2S production has been shown in isolated rabbit [11], human [12], rat [12, 13], and mice corpus cavernosum tissues [16, 58]. In human corpus cavernosum tissue, the expression of protein and mRNA for CBS and CSE is determined, and L-cysteine/H_2S pathway regulates erectile function [12]. Also, CSE is present in peripheral nerves, and CBS/CSE is located in bundles of muscular tissue in trabeculae and vascular smooth muscle cells of human penile arteries [12]. In rat corpus cavernosum tissue, only CSE mRNA and protein have been expressed [13]. In mice corpus cavernosum, the protein expression of CBS, CSE, and 3-MST has found, and H_2S mainly is derived from smooth muscle [59]. L-cysteine-produced endogenous H_2S production partially was inhibited by PAG but not AOAA in rat corpus cavernosum homogenates [13]. The intracavernosal injection of H_2S donors NaHS and Na_2S increased the intracavernosal pressure in rat corpus cavernosum [15]. Moreover, L-cysteine and exogenous H_2S produce relaxation in isolated rabbit [11], human [12], rat [12, 13], and mice cavernosal tissues [16, 58]. CBS inhibitor AOAA causes reduction in L-cysteine-induced relaxations, and CSE inhibitor PAG and AOAA increase the contractile responses to electrical stimulation in human cavernosal strips [12], and L-cysteine-induced relaxation is decreased by PAG in mouse corpus cavernosal [16]. Srilatha and coworkers observed the decreased endogenous H_2S levels in aging rats and suggest that endogenous H_2S levels may be a hallmark for erectile impairment with aging [60]. Besides, diminished H_2S levels due to suppressed expression of CBS and CSE have been reported in corpus cavernosum tissue of hypertensive and diabetic rats, suggesting that defect in H_2S levels might be associated with hypertension-induced [61] or diabetes-induced erectile dysfunction [62]. Reduced CBS and CSE expression is also observed in androgen deficiency-induced ED in rats [63]. In diabetic rats with erectile dysfunction, endogenous H_2S production of penile tissue is decreased due to downregulated expression of the CAT/3-MST and DAO/3-MST and low activities of CBS and CSE [64]. The mechanisms of H_2S regulating penile erection may provide a new therapeutic approach in erectile dysfunction [17].

1.4 The Relaxant Mechanisms of Hydrogen Sulfide in Corpus Cavernosum

In this chapter the mechanisms of H_2S-induced relaxations in corpus cavernosum tissue in consideration of existing literature are discussed below. The possible mechanisms for the relaxant effect of H_2S in corpus cavernosal tissue are summarized schematically in Fig. 2.

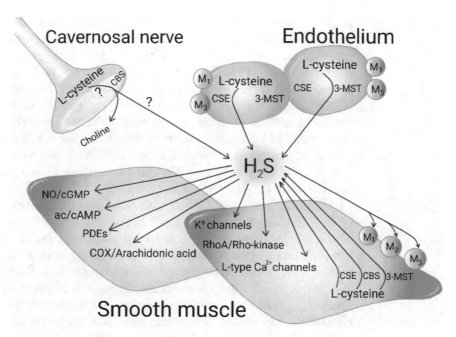

Fig. 2 Schematic representation of synthesis and possible smooth muscle relaxant mechanisms of H₂S in corpus cavernosum

1.4.1 The Contribution of Endothelium

The cavernous endothelium has a critical role in regulating the tone of the underlying smooth muscle and penile erection [20]. H₂S is synthesized endogenously from L-cysteine through CSE enzyme in an endothelium-dependent manner in mice corpus cavernosum tissue [16]. CSE is expressed in vascular tissues, including human umbilical vein, bovine aorta, and mice arterial endothelial cells, and the source of endogenous H₂S is endothelium in these tissues [65]. Exogenous H₂S exhibits its relaxant effect directly through smooth muscle and endothelium-independent manner in mice corpus cavernosum [16]. Consistent with this, the endothelium is not involved in exogenous H₂S- and L-cysteine-induced relaxations in human corpus cavernosum [12].

1.4.2 The Contribution of Nitric Oxide/cGMP Pathway

It is well established that NO released from nitrergic nerves and sinusoidal endothelium is a major endogenous mediator that mediates penile erection [22, 66]. NO exerts its effect in erectile function through the stimulating guanylyl cyclase/cyclic guanosine monophosphate (GC/cGMP) signal transduction system [67]. The cavernous trabecular smooth muscle is considered an extension of the systemic vasculature [29]. In vasculature, it has been suggested that H₂S-induced relaxations are endothelium-dependent and associated with the release of NO or endothelium-derived relaxant factors [3, 68, 69]. Therefore, this possibility has been evaluated in animal and human cavernosal tissues. The higher concentration of exogenous H₂S-induced relaxations was reduced in the presence of nitric oxide synthase (NOS) enzyme inhibitor NG-nitro-L-arginine methyl ester (L-NAME) in human corpus cavernosum strips with endothelium but not endothelium-denuded strips [12]. Exogenous H_2S increased NO production through eNOS and produced proerectile effect by augmenting NO pathway in rat corpus cavernosum, suggesting H_2S may be useful in erectile dysfunction, especially due to endothelial NO deficiency [14]. In mice corpus cavernosum tissue, relaxations to higher concentrations of exogenous H_2S were increased in the presence of NOS inhibitor Nω-nitro-L-arginine (L-NA) and the combination of guanylyl cyclase inhibitor 1H-[1,2,4]oxadiazolo [4,3]quinoxalin-1-one (ODQ) with L-NA potentiated the increase of H_2S-induced relaxations. Authors suggest the inhibitory effect of NO/cGMP might be related to direct chemical reactions between endogenous NO and H_2S and/or depend on the NO/cGMP level in mouse corpus cavernosal tissue [16]. Endogenous H_2S has a role for erectile function through facilitation of nerve-mediated penile tumescence in rat corpus cavernosum [10]. The contractile response to electrical stimulation was increased by CSE inhibitor in human corpus cavernosum [12]. In vivo and in vitro studies have been shown that the inhibition of CSE by PAG potentiated NO-mediated NANC relaxations [13] and enhanced the erectile response to cavernosal nerve stimulation [15]. Ghasemi et al. [13] suggested this increase might be related to direct inhibitory effect of H_2S on NOS activity or a chemical reaction between NO [13]. Recently, it has been reported endogenous H_2S compensates for NO deficiency because of the CSE and 3-MST corpus cavernosum is upregulated in eNOS knockout mice [59]. On the other hand, the erectile response to Na_2S is not altered by L-NAME and is independent of the intact innervation of NO rat corpus cavernosum [15]. Also, in rabbit corpus cavernosum tissue, nitrergic relaxations are independent from endogenous H_2S production [11].

1.4.3 The Contribution of Adenylyl Cyclase/Cyclic Adenosine Monophosphate Pathway

Like in other smooth muscles, corporal smooth muscle relaxation is mediated via the intracellular cyclic nucleotide/protein kinase systems. The activation of adenylyl cyclase (ac) by agonist via specific receptors produces cyclic adenosine monophosphate (cAMP) and results in smooth muscle relaxation [22]. Exogenous H_2S-induced relaxations were reduced by adenylyl cyclase inhibitors in rabbit [11], rat [13], and mice corpus cavernosum tissues [16]. Phosphodiesterase (PDE) enzymes degrade cyclic nucleotides cGMP and cAMP to 5'GMP and 5'AMP [70]. The intracellular cGMP level is regulated by the balance between guanylyl cyclase and PDE [71, 72]. The possible role of PDE in exogenous H_2S-induced relaxations has been investigated in mice corpus cavernosum tissue. PDEV/I inhibitor zaprinast or PDEV inhibitor sildenafil reduced relaxations to H_2S in mice corpus cavernosum tissue, suggesting H_2S may play an inhibitory role on PDE activity, and exhibits its relaxant activity at least in part by inhibiting PDEV in mouse corpus cavernosum [16]. Also, the inhibitory effect of cAMP-specific PDE enzyme blocker has been observed on exogenous H_2S-induced relaxation in mice corpus cavernosum tissue [73]. On the other hand, the erectile responses to nerve and NO donor were not altered by intracavernosal injection of H_2S donor Na_2S in rat corpus cavernosum, suggesting exogenous H_2S donors are not acting to inhibit PDEV in rat corpus cavernosum [15].

1.4.4 The Contribution of K^+ Channels

The K^+ channels play an important role in the H_2S-induced vasodilation in vascular tissues. The role of K^+ channels has been shown in mice [16] and human [12]. It is well known that the ATP-dependent K^+ (K_{ATP}) channel is present in many tissues, including corpus cavernosal smooth muscle. The higher concentration of H_2S-induced relaxations is partially related to K_{ATP} channel in human [12], rat [13], and mice [16] corpus cavernosum tissues. On the other hand, the increased intracavernosal pressure by H_2S donor Na_2S was not reduced in the presence of glibenclamide, a K_{ATP} channel inhibitor, in rat corpus cavernosum [15]. The authors suggest the increase in intracavernosal pressure in response to H_2S donors is not mediated by activation of K_{ATP} channel in the rat [15]. Voltage-gated (K_V) and inward rectifier (K_{IR}) potassium channels are involved in relaxations in mice corpus cavernosum induced by higher concentrations of exogenous H_2S [16]. Exogenous H_2S-induced relaxations are independent from small conductance Ca^{2+}-dependent K^+ (SK_{Ca}^{2+}) channels and intermediate/ large conductance Ca^{2+}-dependent K^+ ($IK_{Ca}^{2+}/BK_{Ca}^{2+}$) channels in mice corpus cavernosum tissue [16]. Distinctively, BK_{Ca}^{2+} channels contribute to mediating erectile response to exogenous H_2S donor Na_2S in rat corpus cavernosum [15].

1.4.5 The Contribution of Voltage-Gated L-Type Ca^{2+} Channels and Na^+-K^+-ATPase

The L-type voltage-gated Ca^{2+} channels are involved in exogenous H_2S-induced relaxations in mouse corpus cavernosum tissue [16]. The contractility of human corpus cavernosum is modulated by the activation of Na^+-K^+-ATPase [74]. There is no contribution of Na^+-K^+-ATPase to mediating relaxant response to exogenous H_2S in mice corpus cavernosum [16].

1.4.6 The Contribution of Muscarinic Acetylcholine Receptors

Even though there are studies indicating the interaction between cholinergic system and H_2S in the vascular [65] and urogenital smooth muscle [75], no study has been performed in penile corpus cavernosal tissue. The first evidence, indicating the contribution of muscarinic receptors has been published in 2016 by Aydinoglu and Ogulener who described the inhibitory effects of atropine, nonselective muscarinic receptor antagonist on exogenous H_2S-induced relaxation [16]. Recently, the same study group clearly showed that selective M_1 inhibitor pirenzepine, M_2 receptor inhibitor AFD X116, and selective M_3 receptor inhibitor 4-DAMP decreased both L-cysteine and exogenous H_2S-induced relaxation in mouse corpus cavernosum tissue, suggesting muscarinic receptor subtypes are involved to H_2S-induced relaxations in mouse corpus cavernosum tissue [76].

1.4.7 The Contribution of Arachidonic Acid/ Cyclooxygenase Cascade

There is an evidence for the role of the phospholipase A_2 (PLA_2), cyclooxygenase (COX), and cytochrome P450-dependent arachidonate metabolites in mediating the relaxant response to exogenous H_2S in mice corpus cavernosum [73]. By contrast, arachidonic acid metabolites do not contribute to the increase in intracavernosal pressure in response to Na_2S in rat corpus cavernosum [15].

1.4.8 The Contribution of RhoA/Rho-Kinase Pathway

The importance of Rho-kinase activity in the maintenance of corporal vasoconstriction and penile *detumescence* has been reported [77–79]. However, it is not extensively clarified what is the role RhoA/Rho-kinase pathway in H_2S-induced relaxations in corpus cavernosum tissue. RhoA, ROCK-1, and ROCK-2 are expressed in mouse corpus cavernosum, and RhoA/Rho-kinase pathway is involved to contractions induced by adrenergic α_1 receptor agonist in this tissue [29]. Most recently, in mouse corpus cavernosum, RhoA/Rho-kinase pathway is partially contributed to relaxant response to exogenous H_2S [73].

1.5 H_2S: A Therapeutic Target for Erectile Dysfunction

The erectile dysfunction may result from vascular, neurologic, psychological, and hormonal factors. Erectile dysfunction is predominately a disease of vascular origin, and vascular risk factors such as aging, diabetes, hypertension, and hypercholesterolemia may increase the incidence of erectile dysfunction [20, 22]. Erectile dysfunction is associated with damaged neurogenic- and endothelium-dependent vasodilator mechanism in corpus cavernosum [25]. Therefore, the restoration of the endothelial function may be considered as the therapeutic strategy for treatment of erectile dysfunction by pharmacological and gene therapy [20]. NO/cGMP pathway is the major mechanism that is involved in erectile function. PDEV inhibitors are used for the treatment of erectile dysfunction. However, PDEV inhibitors are not sufficient in some patients with NO deficiency due to endothelial and neurological damage [28, 80]. Shukla et al. reported that hydrogen sulfide-donating sildenafil (ACS6) has proerectile effect through inhibition of oxidative stress and downregulation of PDEV in cavernosal smooth muscle cell [81]. On the other hand, Meng et al. suggested that H_2S may be useful for the treatment of erectile dysfunction in patients with reduced endothelial NO formation [14]. Recently, it has been suggested that the formation of H_2S

compensates for nitric oxide deficiency and may be a potential therapeutic value in erectile dysfunction [59]. It has been suggested that strategies that increase H_2S formation in penile tissue may be useful in the treatment of erectile dysfunction when NO bioavailability, K_{ATP} channel function, and response to PGE_1 are impaired [15].

Erectile mechanisms have been investigated by in vitro/in vivo animal and human studies [30, 67, 82–85]. In vitro studies provide the evaluation of the isolated animal and human corpus cavernosum tone [30, 67, 82]. Also, the alteration of intracavernosal pressure is determined by in vivo studies [83, 86]. The mouse is a suitable model for investigation of erectile function in vivo and in vitro studies [83]. The in vitro experimental procedure for mice corpus cavernosum is explained below.

2 Materials

1. Isolated organ bath system.

2. Krebs solution: 119 mM NaCl, 4.5 mM KCl, 1.5 mM $CaCl_2$, 2.6 mM $MgCl_2$, 35.1 NaH_2PO_4, 15 mM $NaHCO_3$, 6.7 mM glucose.

3. Phenylephrine.

4. Isotonic force transducer.

5. Recording instrument (polygraph).

3 Methods

1. Kill the mice by cervical dislocation.

2. Remove penis from the mice by dissecting proximally to the penis' ischial attachments and separating the penile crura from the pubic bone, and place it in a Petri dish containing Krebs solution.

3. Excise the glans penis and urethra, and cut fibrous septum between the two corpus cavernosum strips.

4. Dissect from the adherent tissues carefully, keeping the tunica albuginea intact for each corpus cavernosum (0.3 × 0.3 × 4 mm).

5. Tie up the isolated strips at each end with cotton threads (Fig. 3).

6. Mount the cavernosal strips under 0.2 g tension in an organ bath (10 mL) containing Krebs solution.

7. Kept the bath medium at 37 °C and gas with 5% CO_2 and 95% O_2.

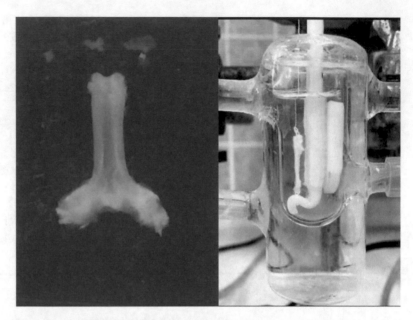

Fig. 3 Isolation of mouse corpus cavernosal tissue and mount in the organ bath

8. Allow the tissue strips to equilibrate for a period of 60 min, and replace the bath solution every 15 min in this period.

9. Record the responses with isotonic transducer on a recorder.

4   Notes

1. Freshly prepared solutions should be ready before the study starts. Krebs solution can be kept +4 °C degree for a day.

2. It is particularly important to allow all the electronic equipment and bath thermostat to warm up before using it.

References

1. Wang R (2002) Two's company, three's a crowd: can H2S be the third endogenous gaseous transmitter? FASEB J 16(13):1792–1798

2. Hosoki R, Matsuki N, Kimura H (1997) The possible role of hydrogen sulfide as an endogenous smooth muscle relaxant in synergy with nitric oxide. Biochem Biophys Res Commun 237(3):527–531

3. Zhao W, Zhang J, Lu Y, Wang R (2001) The vasorelaxant effect of H(2)S as a novel endogenous gaseous K(ATP) channel opener. EMBO J 20(21):6008–6016

4. Cheng Y, Ndisang JF, Tang G, Cao K, Wang R (2004) Hydrogen sulfide-induced relaxation of resistance mesenteric artery beds of rats. Am J

Physiol Heart Circ Physiol 287(5): H2316–H2323

5. Dhaese I, Lefebvre RA (2009) Myosin light chain phosphatase activation is involved in the hydrogen sulfide-induced relaxation in mouse gastric fundus. Eur J Pharmacol 606 (1–3):180–186

6. Dhaese I, Van Colen I, Lefebvre RA (2010) Mechanisms of action of hydrogen sulfide in relaxation of mouse distal colonic smooth muscle. Eur J Pharmacol 628(1–3):179–186

7. Fusco F, di Villa Bianca R, Mitidieri E, Cirino G, Sorrentino R, Mirone V (2012) Sildenafil effect on the human bladder involves the L-cysteine/hydrogen sulfide pathway: a

novel mechanism of action of phosphodiesterase type 5 inhibitors. Eur Urol 62 (6):1174–1180

8. Rashid S, Heer JK, Garle MJ, Alexander SP, Roberts RE (2013) Hydrogen sulphide-induced relaxation of porcine peripheral bronchioles. Br J Pharmacol 168 (8):1902–1910

9. Dunn WR, Alexander SP, Ralevic V, Roberts RE (2016) Effects of hydrogen sulphide in smooth muscle. Pharmacol Ther 158:101–113

10. Srilatha B, Adaikan PG, Moore PK (2006) Possible role for the novel gasotransmitter hydrogen sulphide in erectile dysfunction—a pilot study. Eur J Pharmacol 535 (1–3):280–282

11. Srilatha B, Adaikan PG, Li L, Moore PK (2007) Hydrogen sulphide: a novel endogenous gasotransmitter facilitates erectile function. J Sex Med 4(5):1304–1311

12. d'Emmanuele di Villa Bianca R, Sorrentino R, Maffia P, Mirone V, Imbimbo C, Fusco F, De Palma R, Ignarro LJ, Cirino G (2009) Hydrogen sulfide as a mediator of human corpus cavernosum smooth-muscle relaxation. Proc Natl Acad Sci U S A 106(11):4513–4518

13. Ghasemi M, Dehpour AR, Moore KP, Mani AR (2012) Role of endogenous hydrogen sulfide in neurogenic relaxation of rat corpus cavernosum. Biochem Pharmacol 83 (9):1261–1268

14. Meng J, Ganesan Adaikan P, Srilatha B (2013) Hydrogen sulfide promotes nitric oxide production in corpus cavernosum by enhancing expression of endothelial nitric oxide synthase. Int J Impot Res 25(3):86–90

15. Jupiter RC, Yoo D, Pankey EA, Reddy VV, Edward JA, Polhemus DJ, Peak TC, Katakam P, Kadowitz PJ (2015) Analysis of erectile responses to H2S donors in the anesthetized rat. Am J Physiol Heart Circ Physiol 309(5):H835–H843

16. Aydinoglu F, Ogulener N (2016) Characterization of relaxant mechanism of H2S in mouse corpus cavernosum. Clin Exp Pharmacol Physiol 43(4):503–511

17. Huang YM, Cheng Y, Jiang R (2012) Hydrogen sulfide and penile erection. Zhonghua Nan Ke Xue 18:823–826

18. Andersson KE, Wagner G (1995) Physiology of penile erection. Physiol Rev 75(1):191–236

19. Ralph DJ (2005) Normal erectile function. Clin Cornerstone 7(1):13–18

20. Bivalacqua TJ, Usta MF, Champion HC, Kadowitz PJ, Hellstrom WJ (2003) Endothelial dysfunction in erectile dysfunction: role of the endothelium in erectile physiology and disease. J Androl 24(6 Suppl):S17–S37

21. Burnett AL (2006) The role of nitric oxide in erectile dysfunction: implications for medical therapy. J Clin Hypertens (Greenwich) 8 (12):53–62

22. Andersson KE (2001) Pharmacology of penile erection. Pharmacol Rev 53(3):417–450

23. Saenz de Tejada I, Blanco R, Goldstein I, Azadzoi K, de las Morenas A, Krane RJ, Cohen RA (1988) Cholinergic neurotransmission in human corpus cavernosum. I Respons Isolated Tissue J Physiol 254(3 Pt 2): H459–H467

24. Azadzoi KM, Kim N, Brown ML, Goldstein I, Cohen RA, Saenz de Tejada I (1992) Endothelium-derived nitric oxide and cyclooxygenase products modulate corpus cavernosum smooth muscle tone. J Urol 147(1):220–225

25. Simonsen U, García-Sacristán A, Prieto D (2002) Penile arteries and erection. J Vasc Res 39(4):283–303

26. Hanyu S, Iwanaga T, Kano K, Sato S (1987) Mechanism of penile erection in the dog. Pressure-flow study combined with morphological observation of vascular casts. Urol Int. 42(6):401–412

27. Fazio L, Broc G (2004) Erectile dysfunction: management update. CMAJ 170(9):1429

28. Andersson KE (2011) Mechanisms of penile erection and basis for pharmacological treatment of erectile dysfunction. Pharmacol Rev 63(4):811–859

29. Kumcu EK, Aydinoglu F, Astarci E, Ogulener N (2016) The effect of sub-chronic systemic ethanol treatment on corpus cavernosal smooth muscle contraction: the contribution of RhoA/Rho-kinase. Naunyn Schmiedeberg's Arch Pharmacol 389(3):249–258

30. Pickard RS, King P, Zar MA, Powell PH (1994) Corpus cavernosal relaxation in impotent men. Br J Urol 74(4):485–491

31. Kamaoun P (2004) Endogenous production of hydrogen sulfide in mammals. Amino Acids 26 (3):243–254

32. Łowicka E, Bełtowski J (2007) Hydrogen sulfide (H2S)—the third gas of interest for pharmacologists. Pharmacol Rep 59(1):4–24

33. Mancardi D, Penna C, Merlino A, Del Soldato P, Wink DA, Pagliaro P (2009) Physiological and pharmacological features of the novel gasotransmitter: hydrogen sulfide. Biochim Biophys Acta 1787(7):864–872

34. Martin GR, McKnight GW, Dicay MS, Coffin CS, Ferraz JG, Wallace JL (2010) Hydrogen sulphide synthesis in the rat and mouse gastrointestinal tract. Dig Liver Dis 42(2):103–109

35. Whiteman M, Le Trionnaire S, Chopra M, Fox B, Whatmore J (2011) Emerging role of hydrogen sulfide in health and disease: critical appraisal of biomarkers and pharmacological tools. Clin Sci (Lond) 121(11):459–488

36. Paul BD, Snyder SH (2012) H2S signalling through protein sulfhydration and beyond. Nat Rev Mol Cell Biol 13(8):499–507

37. Doeller JE, Isbell TS, Benavides G, Koenitzer J, Patel H, Patel RP, Lancaster JR Jr, Darley-Usmar VM, Kraus DW (2005) Polarographic measurement of hydrogen sulfide production and consumption by mammalian tissues. Anal Biochem 341(1):40–51

38. Patel P, Vatish M, Heptinstall J, Wang R, Carson RJ (2009) The endogenous production of hydrogen sulphide in intrauterine tissues. Reprod Biol Endocrinol 7:10

39. Fang L, Zhao J, Chen Y, Ma T, Xu G, Tang C, Liu X, Geng B (2009) Hydrogen sulfide derived from periadventitial adipose tissue is a vasodilator. J Hypertens 27(11):2174–2185

40. Du JT, Li W, Yang JY, Tang CS, Li Q, Jin HF (2013) Hydrogen sulfide is endogenously generated in rat skeletal muscle and exerts a protective effect against oxidative stress. Chin Med J 126(5):930–936

41. Abe K, Kimura H (1996) The possible role of hydrogen sulfide as an endogenous neuromodulator. J Neurosci 16(3):1066–1071

42. Pae HO, Lee YC, Jo EK, Chung HT (2009) Subtle interplay of endogenous bioactive gases (NO, CO and H(2)S) in inflammation. Arch Pharm Res 32(8):1155–1162

43. Munaron L, Avanzato D, Moccia F, Mancardi D (2013) Hydrogen sulfide as a regulator of calcium channels. Cell Calcium 53(2):77–84

44. Kimura H (2010) Hydrogen sulfide: from brain to gut. Antioxid Redox Signal 12(9):1111–1123

45. Kimura H (2011) Hydrogen sulfide: its production, release and functions. Amino Acids 41(1):113–121

46. Szabó C (2007) Hydrogen sulphide and its therapeutic potential. Nat Rev Drug 6(11):917–935

47. Kabil O, Vitvitsky V, Xie P, Banerjee R (2011) The quantitative significance of the transsulfuration enzymes for H2S production in murine tissues. Antioxid Redox Signal 15(2):363–372

48. Al-Magableh MR, Hart JL (2011) Mechanism of vasorelaxation and role of endogenous hydrogen sulfide production in mouse aorta. Naunyn Schmiedeberg's Arch Pharmacol 383(4):403–413

49. Kaneko Y, Kimura Y, Kimura H, Niki I (2006) L-cysteine inhibits insulin release from the pancreatic beta-cell: possible involvement of metabolic production of hydrogen sulfide, a novel gasotransmitter. Diabetes 55(5):1391–1397

50. d'Emmanuele di Villa Bianca R, Mitidieri E, Fusco F, Russo A, Pagliara V, Tramontano T, Donnarumma E, Mirone V, Cirino G, Russo G, Sorrentino R (2016) Urothelium muscarinic activation phosphorylates CBS (Ser227) via cGMP/PKG pathway causing human bladder relaxation through H2S production. Sci Rep 6:31491

51. Diwakar L, Ravindranath V (2007) Inhibition of cystathionine-gamma-lyase leads to loss of glutathione and aggravation of mitochondrial dysfunction mediated by excitatory amino acid in the CNS. Neurochem Int 50(2):418–426

52. Nagahara N, Ito T, Kitamura H, Nishino T (1998) Tissue and subcellular distribution of mercaptopyruvate sulfurtransferase in the rat: confocal laser fluorescence and immunoelectron microscopic studies combined with biochemical analysis. Histochem Cell Biol 110(3):243–250

53. Shibuya N, Mikami Y, Kimura Y, Nagahara N, Kimura H (2009) Vascular endothelium expresses 3-mercaptopyruvate sulfurtransferase and produces hydrogen sulfide. J Biochem 146(5):623–626

54. Shibuya N, Tanaka M, Yoshida M, Ogasawara Y, Togawa T, Ishii K, Kimura H (2009) 3-Mercaptopyruvate sulfurtransferase produces hydrogen sulfide and bound sulfane sulfur in the brain. Antioxid Redox Signal 11(4):703–714

55. Shibuya N, Koike S, Tanaka M, Ishigami-Yuasa M, Kimura Y, Ogasawara Y, Fukui K, Nagahara N, Kimura H (2013) A novel pathway for the production of hydrogen sulfide from D-cysteine in mammalian cells. Nat Commun 4:1366

56. Kimura H (2014) The physiological role of hydrogen sulfide and beyond. Nitric Oxide 41:4–10

57. Martelli A, Testai L, Breschi MC, Lawson K, McKay NG, Miceli F, Taglialatela M, Calderone V (2013) Vasorelaxation by hydrogen sulphide involves activation of Kv7 potassium channels. Pharmacol Res 70(1):27–34

58. Yetik-Anacak G, Dereli MV, Sevin G, Ozzayım O, Erac Y, Ahmed A (2015) Resveratrol stimulates hydrogen sulfide (H2 S) formation to relax murine corpus cavernosum. J Sex Med 12(10):2004–2012

59. Yetik-Anacak G, Dikmen A, Coletta C, Mitidieri E, Dereli M, Donnarumma E,

d'Emmanuele di Villa Bianca R, Sorrentino R (2016) Hydrogen sulfide compensates nitric oxide deficiency in murine corpus cavernosum. Pharmacol Res 113(Pt A):38–43

60. Srilatha B, Muthulakshmi P, Adaikan PG, Moore PK (2012) Endogenous hydrogen sulfide insufficiency as a predictor of sexual dysfunction in aging rats. Aging Male 15 (3):153–158

61. Zhu XB, Jiang J, Jiang R, Chen F (2014) Expressions of CSE and CBS in the corpus cavernosum of spontaneous hypertensive rats. Zhonghua Nan Ke Xue 20(1):4–9

62. Huang YM, Xia JY, Jiang R (2014) Expressions of CSE and CBS in the penile corpus cavernosum of hyperglycemia rats and their implications. Zhonghua Nan Ke Xue 20(4):299–303

63. Zuo C, Huang YM, Jiang R, Yang HF, Cheng B, Chen F (2014) Endogenous hydrogen sulfide and androgen deficiency-induced erectile dysfunction in rats. Zhonghua Nan Ke Xue 20(7):605–612

64. Zhang Y, Yang J, Wang T, Wang SG, Liu JH, Yin CP, Ye ZQ (2016) Decreased endogenous hydrogen sulfide generation in penile tissues of diabetic rats with erectile dysfunction. J Sex Med 13(3):350–360

65. Yang G, Wu L, Jiang B, Yang W, Qi J, Cao K, Meng Q, Mustafa AK, Mu W, Zhang S, Snyder SH, Wang R (2008) H2S as a physiologic vasorelaxant: hypertension in mice with deletion of cystathionine gamma-lyase. Science 322 (5901):587–590

66. Kim N, Azadzoi KM, Goldstein I, Saenz de Tejada I (1991) A nitric oxide-like factor mediates nonadrenergic-noncholinergic neurogenic relaxation of penile corpus cavernosum smooth muscle. J Clin Invest 88(1):112–118

67. Hedlund P, Aszodi A, Pfeifer A, Alm P, Hofmann F, Ahmad M, Fassler R, Andersson KE (2000) Erectile dysfunction in cyclic GMP-dependent kinase I-deficient mice. Proc Natl Acad Sci U S A 97(5):2349–2354

68. Zhao W, Wang R (2002) H(2)S-induced vasorelaxation and underlying cellular and molecular mechanisms. Am J Physiol Heart Circ Physiol 283(2):H474–H480

69. Wang YF, Mainali P, Tang CS, Shi L, Zhang CY, Yan H, Liu XQ, Du JB (2008) Effects of nitric oxide and hydrogen sulfide on the relaxation of pulmonary arteries in rats. Chin Med J 121(5):420–423

70. Lugnier C (2006) Cyclic nucleotide phosphodiesterase (PDE) superfamily: a new target for the development of specific therapeutic agents. Pharmacol Ther 109(2):366–398

71. Lucas KA, Pitari GM, Kazerounian S, Ruiz-Stewart I, Park J, Schulz S, Chepenik KP, Waldman SA (2000) Guanylyl cyclases and signaling by cyclic GMP. Pharmacol Rev 52(3):375–414

72. Essayan DM (2001) Cyclic nucleotide phosphodiesterases. J Allergy Clin Immunol 108 (5):671–680

73. Aydinoglu F, Ogulener N (2016) The effects of cyclooxygenase, nitric oxide, phosphodiesterase IV and Rho-kinase inhibitors on hydrogen sulfide-induced relaxant response in mouse corpus cavernosum. Paper presented at the 7th European Congress of Pharmacology, Istanbul, 26–30 June 2016

74. Gupta S, Moreland RB, Munarriz R, Daley J, Goldstein I, Saenz de Tejada I (1995) Possible role of Na(+)-K(+)-ATPase in the regulation of human corpus cavernosum smooth muscle contractility by nitric oxide. Br J Pharmacol 116(4):2201–2206

75. Fernandes VS, Xin W, Petkov GV. Novel mechanism of hydrogen sulfide-induced guinea pig urinary bladder smooth muscle contraction: role of BK channels and cholinergic neurotransmission. Am J Physiol Cell Physiol 309 (2):C107–16

76. Dalkir FT, Aydinoglu F, Ogulener N (2016) The interaction of L-cysteine/hydrogen sulfide pathway and muscarinic acetylcholine receptors (mAChRs) in Mouse Corpus Cavernosum. Paper presented at the 7th European Congress of Pharmacology, Istanbul, 26–30 June 2016

77. Chitaley K, Wingard CJ, Clinton Webb R, Branam H, Stopper VS, Lewis RW, Mills TM (2001) Antagonism of Rho-kinase stimulates rat penile erection via a nitric oxide-independent pathway. Nat Med 7:119–122

78. Rees RW, Ralph DJ, Royle M, Moncada S, Cellek S (2001) Y-27632, an inhibitor of Rho-kinase, antagonizes noradrenergic contractions in the rabbit and human penile corpus cavernosum. Br J Pharmacol 133:455–458

79. Waldkirch E, Uckert S, Yildirim H, Sohn M, Jonas U, Stief CG, Andersson KE, Hedlund P (2005) Cyclic AMP-specific and cyclic GMP-specific phosphodiesterase isoenzymes in human cavernous arteries—immunohistochemical distribution and functional significance. World J Urol 23(6):405–410

80. Sáenz de Tejada I, Angulo J, Cellek S, González-Cadavid N, Heaton J, Pickard R, Simonsen U (2004) Physiology of erectile function. J Sex Med 1(3):254–265

81. Shukla N, Rossoni G, Hotston M, Sparatore A, Del Soldato P, Tazzari V, Persad R, Angelini GD, Jeremy JY (2009) Effect of hydrogen sulphide-donating sildenafil (ACS6) on erectile

function and oxidative stress in rabbit isolated corpus cavernosum and in hypertensive rats. BJU Int 103:1522–1529

82. Gocmen C, Uçar P, Singirik E, Dikmen A, Baysal F (1997) An in vitro study of nonadrenergic-noncholinergic activity on the cavernous tissue of mouse. Urol Res 25 (4):269–275

83. Mizusawa H, Hedlund P, Håkansson A, Alm P, Andersson KE (2001) Morphological and functional in vitro and in vivo characterization of the mouse corpus cavernosum. Br J Pharmacol 132(6):1333–1341

84. Baracat JS, Teixeira CE, Okuyama CE, Priviero FB, Faro R, Antunes E, De Nucci G (2003) Relaxing effects induced by the soluble guanylyl cyclase stimulator BAY 41-2272 in human and rabbit corpus cavernosum. Eur J Pharmacol 477(2):163–169

85. Ayajiki K, Hayashida H, Tawa M, Okamura T, Toda N (2009) Characterization of nitrergic function in monkey penile erection in vivo and in vitro. Hypertens Res 32(8):685–689

86. Angulo J, Peiró C, Sanchez-Ferrer CF, Gabancho S, Cuevas P, Gupta S, Sáenz de Tejada I (2001) Differential effects of serotonin reuptake inhibitors on erectile responses, NO-production, and neuronal NO synthase expression in rat corpus cavernosum tissue. Br J Pharmacol 134(6):1190–1194

Chapter 11

Pharmacological Tools for the Study of H₂S Contribution to Angiogenesis

Lucia Morbidelli, Martina Monti, and Erika Terzuoli

Abstract

Recently, hydrogen sulfide (H_2S) has been characterized as an endogenous mediator able to control a series of cellular and tissue functions relevant for tissue homeostasis and repair such as angiogenesis. This chapter describes the tools and their use in a set of angiogenesis assays performed by using cultured endothelial cells in order to study the relevance of exogenous or endogenous H_2S production and release during the occurrence of angiogenesis.

Key words Angiogenesis, Endothelial cells, Hydrogen sulfide, Cystathionine γ-lyase, Proliferation, Migration, Chemoinvasion, Permeability, Gelatinase

1 Introduction

Angiogenesis, the process by which new blood vessels are formed from pre-existing ones, plays a crucial role both in physiological (wound healing, embryonic development) and pathological conditions (diabetic retinopathy, arthritis, tumor growth, and metastasis) [1]. Angiogenesis is a tightly controlled process. In the last decades, different families of regulators of angiogenesis, both stimulators and inhibitors, have been discovered and characterized by a molecular point of view [2].

The steps required for new vessel growth are biologically complex. Sprout formation during the initial steps of the angiogenic process is commonly preceded by strong and persistent vasodilation and increased vascular permeability. Later events include endothelial cell proliferation, migration, and protease release [3].

Vascular endothelium is a metabolically active organ that processes and participates to the synthesis of vasoactive and proangiogenic mediators, among which are gaseous transmitters [4]. We have largely contributed to demonstrate the role of endogenous and exogenous nitric oxide in the control of angiogenesis [5–7]. In the last years, we have focused the attention on hydrogen sulfide

Jerzy Bełtowski (ed.), *Vascular Effects of Hydrogen Sulfide: Methods and Protocols*, Methods in Molecular Biology, vol. 2007, https://doi.org/10.1007/978-1-4939-9528-8_11, © Springer Science+Business Media, LLC, part of Springer Nature 2019

(H$_2$S), demonstrating its increased bioavailability following the exposure of cultured endothelium to the active moiety of the angiotensin-converting enzyme inhibitor zofenopril, zofenoprilat [8, 9].

H$_2$S is a physiologic messenger molecule involved in cardiovascular health. H$_2$S is generated in the periphery by cystathionine γ-lyase (cystathionase; CSE), while in the brain its biosynthesis may involve cystathionine β-synthase (CBS) [10, 11]. CSE knockout (CSE$^{-/-}$) mice display hypertension and a major decrease in endothelial-derived relaxing factor activity, establishing H$_2$S as an important vasorelaxant [12, 13] and proangiogenic molecule [14].

The study of exogenous and endogenous H$_2$S in the control of endothelial functions related to angiogenesis and other endothelial functions relies on the availability of pharmacological and biochemical tools.

2 Materials

Prepare all solutions using ultrapure water (prepared by purifying deionized water to attain a resistivity of 18 MΩ·cm at 25 °C) and analytical grade reagents. Prepare and store all reagents at room temperature (RT) (unless indicated otherwise) and under vertical hood to guarantee sterility. Diligently follow all rules for individual protection (dedicated lab coat, disposable gloves, glasses) and waste disposal regulations when disposing waste materials.

Instruments necessary for all the procedures and protocols are:

1. Vertical hood (biohazard level 2).

2. Disposable pipettes and Pipet-Aid dispenser.

3. Hemocytometer for cell counting.

4. Humidified incubator set at 37 °C with an atmosphere of 5% CO$_2$ in air.

5. Inverted microscope Nikon Eclipse TE 300 (Nikon, Tokyo, Japan) equipped with 4, 10, 20, and 40× objectives.

6. Refrigerated centrifuge.

7. Refrigerator and freezers.

2.1 Pharmacological Tools

1. Solution of NaHS (Sigma-Aldrich, St. Louis, MO, USA) in Dulbecco's phosphate-buffered saline (PBS) with calcium and magnesium (*see* **Note 1**).

2. D,L-propargylglycine (PAG) from Sigma-Aldrich (St. Louis, MO, USA) solubilized in PBS (*see* **Note 2**).

2.2 Endothelial Cells

1. Endothelial cells from human umbilical vein (HUVEC) from a commercial source (*see* **Note 3**) maintained in endothelial cell growth medium 2 (EGM-2, Lonza, Basel, Switzerland), with

the recommended supplements, passaged once reaching the confluence onto gelatin-coated flasks. HUVEC may be used up to passage 8 (*see* **Note 4**).

2. Trypsin/ethylenediaminetetraacetic acid (EDTA) solution from commercial source (Sigma-Aldrich, St. Louis, MO, USA).

3. Tissue culture flasks, Petri dishes, or multiwell plates.

4. Bovine gelatin (Sigma-Aldrich, St. Louis, MO, USA).

5. Gelatin-coated flasks/dishes prepared by adding 1% bovine gelatin solution (*see* **Note 5**) to the flasks/dishes to cover the entire surface and incubating at 37 °C for 15 min. Aspirate gelatin and wash the flasks with sterile PBS to remove gelatin excess.

6. Fetal bovine serum (FBS, Hyclone GE Healthcare Life Sciences, South Logan, UT, USA).

2.3 siRNA-Mediated Gene Silencing for CSE

1. Endothelial cells (e.g., HUVEC).

2. 6-well tissue culture plates.

3. Transfection medium: endothelial basal medium (EBM, Lonza Euroclone S.p.a, Pero, MI, Italy) without antibiotics.

4. Gelatin solution, trypsin-EDTA solution, and serum as described in Subheading 2.2.

5. Graduated pipette (Gilson P20, P200, P1000).

6. Sterile and RNase- and DNase-free 200 and 1000 μL pipette tips for Gilson P20, P200, and P1000 (Euroclone S.p.a, Pero, MI, Italy).

7. RNase- and DNase-free 1500 μL vials (Sarstedt S.r.l., Verona, Italy).

8. Opti-MEM medium (Gibco, Thermo Fisher Scientific, Paisley, UK).

9. Lipofectamine 3000 (Life Technologies, Thermo Fisher Scientific). Lipofectamine 3000 has high efficiency and low toxicity for sensitive cells such as endothelial cells.

10. The siRNA sequence of human CSE (ID: s3712) and the validated silencer negative control from Ambion (Thermo Fisher Scientific). Dilute siRNA duplexes to a concentration of 20 μM in RNase-free water (*see* **Note 6**).

2.4 Fluorimetric Tool for H₂S Detection

1. Endothelial cells (e.g., HUVEC).

2. 24- and 96-well tissue culture plates.

3. Endothelial basal medium (EBM-2, Lonza, Basel, Switzerland).

4. Gelatin solution, trypsin-EDTA solution, and serum as described in Subheading 2.2.

5. Graduated pipette (Gilson P20, P200, P1000).

6. 1 cm round glass coverslips.

7. WSP-1 fluorescent dye (Cayman Chemical, Ann Arbor, MI, USA) (*see* **Note 7**).

8. Dimethyl sulfoxide (DMSO, Sigma-Aldrich, St. Louis, MO, USA).

9. 4% paraformaldehyde (Thermo Fisher Scientific, Prokeme, Calenzano, Florence, Italy).

10. Phosphate-buffered saline (PBS).

11. Infinite® 200 PRO NanoQuant (Thermo Fisher Scientific).

12. British standards microscope slides (Thermo Fisher Scientific).

13. Microscope cover glass (Thermo Fisher Scientific).

14. DAPI and Fluoromount Aqueous Mounting Medium (Sigma-Aldrich, St. Louis, MO, USA).

15. Confocal microscope (Zeiss LSM500).

2.5 Cell Proliferation

1. Endothelial cells (e.g., HUVEC).

2. 96-well tissue culture plates.

3. Endothelial basal medium (EBM-2, Lonza, Basel, Switzerland).

4. Gelatin solution, trypsin-EDTA solution, and serum as described in Subheading 2.2.

5. Graduated pipette (Gilson P20, P200, P1000).

6. Sterile 200 and 1000 μL pipette tips for Gilson P20, P200, and P1000.

7. Diff-Quik reagents (Panreac, Barcelona, Spain) for cell staining.

2.6 Cell Migration (Scratch Assay)

1. Endothelial cells (e.g., HUVEC).

2. 24-well tissue culture plates.

3. Endothelial basal medium (EBM-2, Lonza, Basel, Switzerland).

4. Gelatin solution, trypsin-EDTA solution, and serum as described in Subheading 2.2.

5. Graduated pipette (Gilson P20, P200, P1000).

6. Sterile 200 and 1000 μL pipette tips for Gilson P20, P200, and P1000.

7. Arabinosylcytosine (ARA-C, Sigma-Aldrich, St. Louis, MO, USA), as antimitotic agent.

8. Inverted microscope Nikon Eclipse TE 300 (Nikon, Tokyo, Japan) equipped with 10× objective and digital camera.

9. ImageJ software (NIH free software).

2.7 Cell Chemotaxis and Chemoinvasion

1. Endothelial cells (e.g., HUVEC).

2. 48-well microchemotaxis chamber (or Boyden chambers) (Neuro Probe, Gaithersburg, MD, USA).

3. Polyvinylpyrrolidone (PVP)-free polycarbonate filter, 8 μm pore size (Prokeme, Calcnzano, Florence, Italy).

4. Endothelial basal medium (EBM-2, Lonza, Basel, Switzerland).

5. Gelatin solution, trypsin-EDTA solution, and serum as described in Subheading 2.2.

6. Graduated pipette (Gilson P20, P200, P1000).

7. Sterile 200 and 1000 μL pipette tips for Gilson P20, P200, and P1000.

8. Fibronectin and collagen type I (Sigma-Aldrich St. Louis, MO, USA).

9. Diff-Quik reagents (Panreac, Barcelona, Spain) for cell staining.

10. Histovitrex-Erba Rs Mounting (Carlo Erba Reagents S.A.S. Chaussée du Vexin, France).

11. British standard microscope slides (Thermo Fisher Scientific).

12. Microscope cover glass (Thermo Fisher Scientific).

2.8 Gelatin Zymography

1. Endothelial cells (e.g., HUVEC).

2. 96-well tissue culture plates.

3. Endothelial basal medium (EBM-2, Lonza, Basel, Switzerland).

4. Gelatin solution, trypsin-EDTA solution, and serum as described in Subheading 2.2.

5. Graduated pipette (Gilson P20, P200, P1000).

6. Resolving gel buffer: 1.5 M Tris–HCl, pH 8.8 (Bio-rad, Hercules, CA, USA).

7. Stacking gel buffer: 0.5 M Tris–HCl, pH 6.8 (Bio-rad, Hercules, CA, USA).

8. 40% acrylamide/bis solutions 29:1 (Bio-rad, Hercules, CA, USA).

9. Ammonium persulfate (APS) (Sigma-Aldrich St. Louis, MO, USA): 10% solution in water (*see* **Note 8**).

10. N,N,N,N'-Tetramethylethylenediamine (TEMED) (Sigma-Aldrich St. Louis, MO, USA), stored at 4 °C (*see* **Note 9**).

11. Sodium dodecyl sulfate-polyacrylamide gel electrophoresis (SDS-PAGE) running buffer: 10× Tris/Glycine/SDS Buffer (Bio-rad, Hercules, CA, USA). Prepare 1 L of solution as followed: 100 mL of running buffer and 900 mL of water.

12. 4× Laemmli sample buffer (Bio-rad, Hercules, CA, USA), stored at 4 °C.

13. Protein molecular weight markers (Precision Plus Protein Dual Color Standards, Bio-rad) stored at −20 °C.

14. Washing buffer solution: 2.5% Triton X-100 (Sigma-Aldrich, St. Louis, MO, USA) in water.

15. Developing buffer: 50 mM Tris–HCl (pH 7.4), 5 mM $CaCl_2$, and 200 mM NaCl in water.

16. Coomassie Brilliant Blue solution (0.05% Coomassie Brilliant Blue, 45% methanol, 10% acetic acid, 45% water).

17. Destaining solution (45% methanol, 10% acetic acid, 45% water).

18. Distilled water.

19. Methanol.

20. Mini-protean tetra system, short plates Mini-protean, and outer glass plate 1.5 mm, M-P3 (Bio-rad, Hercules, CA, USA).

21. Power pack 1000 (Bio-rad, Hercules, CA, USA).

22. Chemidoc (Bio-rad, Hercules, CA, USA).

2.9 Matrigel Plug Assay

1. Endothelial cells (e.g., HUVEC).

2. Endothelial basal medium (EBM-2, Lonza, Basel, Switzerland).

3. Trypsin-EDTA solution and serum as described in Subheading 2.2.

4. Matrix Matrigel™ (BD Biosciences, Becton Dickinson, Milan, Italy) (*see* **Note 10**).

5. 24-well tissue culture plates.

6. Gilson pipette tips for P1000 (*see* **Note 11**).

7. Sterilized scissors.

8. Inverted microscope (Microscope Nikon Eclipse TE 300, Nikon, Tokyo, Japan) equipped with 4 and 10× objectives and digital camera.

3 Methods

All procedures are carried out at RT unless otherwise specified and under vertical hood to guarantee sterility.

3.1 Cell Culture Maintenance	The culture of human endothelial cells requires specialized culture media which are indicated by the manufacturer instructions. It's very important to follow the culture indications to ensure cell survival and maintenance of phenotypic pattern. Primary endothelial cells have a finite life span: they maintain their proliferating rate for maximum eight passages, and then they become senescent and stop to proliferate. Always control cell confluency and morphology at the inverted microscope.

3.1.1 Cell Thawing

1. Prepare endothelial growth medium adding media supplements to the basal medium, thawing media supplements rapidly at 37 °C.

2. Add pre-warmed growth medium to the Petri dish pre-coated with gelatin, e.g., 8 mL per dish of 6 cm diameter and 15 mL for dish of 10 cm diameter.

3. Thaw a vial of cryopreserved endothelial cells rapidly at 37 °C, and seed the cells in the previously prepared Petri dish. Put in incubator at 37 °C set at 5% CO_2 in air.

4. After 4 h remove cryoprotectant (dimethyl sulfoxide, DMSO) changing medium with fresh growth medium (5 mL for 6 cm diameter dish and 10 mL for 10 cm diameter dish).

5. Harvest or subculture cells when they reach 90–100% confluence.

3.1.2 Cell Subculture

1. Prepare gelatin pre-coated dishes.

2. Remove medium from confluent dishes.

3. Rinse the cells with 2 mL trypsin-EDTA solution (for 10 cm diameter Petri dishes).

4. Add 2 mL trypsin-EDTA to cover the entire surface of the cell monolayer.

5. After about 2 min, examine the cells microscopically to monitor detachment.

6. Collect the cells by aspirating the cell solution using a disposable pipette and transferring to a suitable sized vial containing 700 μL of serum (*see* **Note 12**).

7. After centrifugation, discard the supernatant and resuspend the cells in growth medium.

8. Reseed the cells at the appropriate density as requested by the experimental protocol.

9. For routine culture maintenance, dilute cells 1:3 twice a week.

3.2 Endothelial Cell Silencing of CSE

1. Detach the cells as reported in the Subheading 3.1.2.

2. Count the cells using a hemocytometer.

3. Seed cells at the density of 1.8×10^5/well in 6-well plates in 10% FBS medium (final volume of each well 3 mL).

4. Allow cell adherence in incubator at 37 °C for 24 h.

5. Day 2: Replace the medium with fresh transfection medium (1.5 mL/well).

6. Day 3: Transfect HUVEC with siRNA oligonucleotide using Lipofectamine 3000. For transfection in a 6-well plate, dilute 3 μL of siRNA oligonucleotide stock with 107 μL Opti-MEM in a RNase-free microfuge tube. Dilute 3 μL of Lipofectamine 3000 with 107 μL Opti-MEM in another RNase-free microfuge tube.

7. Combine the two solutions and incubate for 7 min at RT.

8. Add the siRNA/lipid mix dropwise on the cells while swirling the plate gently.

9. Incubate for 4 h at 37 °C.

10. Add 110 μL of FBS to each well.

11. Assay cells for functional/biochemical assessment 48 h after transfection. Western blot analysis for CSE is necessary to validate the silencing.

3.3 Fluorimetric Assay for H₂S Production in Endothelial Cells

WSP-1-related fluorescence can detect both H_2S released in the cell supernatant and produced inside cells. These data are necessary to assess the activity of novel H_2S donors and to correlate the endogenous production of the gaseous transmitter with angiogenesis, evaluated in the functional assays [9].

1. Harvest cells from Petri dish as described in Subheading 3.2, and plate on 1 cm round glass coverslips in 24-well plates at 2.0×10^4 cells/well in 10% FBS medium (final volume each well 1 mL).

2. Incubate at 37 °C in a humidified incubator at 5% CO_2 in air.

3. Allow cell adherence in incubator at 37 °C for 24 h.

4. Prepare appropriate H_2S donors and metabolic inhibitors and keep on ice.

5. Remove media and treat cells for time ranging from minutes to 6 h with H_2S donors (1–100 μM) or other stimuli in the absence or presence of PAG (3 mM) or other metabolic inhibitors. Perform these treatments in EBM-2 with low serum concentration (0.1%) in a final volume of 500 μL. Repeat each experimental point in triplicate.

6. Incubate the cells with WSP-1 in the dark at final concentration of 100 μM (*see* **Note 13**).

7. Read fluorescence (excitation 485 nm, emission 535 nm) every 7 min up to 35 min. Collect fluorescence data by means of

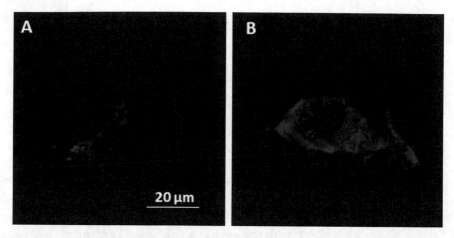

Fig. 1 Representative confocal images of intracellular WSP-1 related fluorescence after 6 h incubation in control condition (0.1% serum) and following Zofenoprilat (10 μM). ×63 magnification; scale bar 20 μm

multiwell plate reader Tecan (Thermo Fisher Scientific, Waltham, USA).

8. Fix cells in cold acetone for 10 min at RT.

9. Wash three times with PBS (5 min each).

10. Incubate with DAPI (1 μg/mL) for 10 min to counterstain the nuclei.

11. Wash cells three times with PBS.

12. Mount coverslips and let dry at RT overnight.

13. Reveal cell-related fluorescence by confocal microscopy (Zeiss LSM500, Milan, Italy) as in Fig. 1.

3.4 Cell Proliferation

Cell number can be evaluated following incubation with H₂S donors or other stimuli in the absence/presence of CSE or other metabolic inhibitors to assess the relevance of exogenous or endogenous H₂S in the promotion of cell proliferation [8, 9].

1. Detach cells as reported in the Subheading 3.1.2.

2. Count the cells using a hemocytometer.

3. Seed cells at the density of 1.5×10^3 cells/well in 96 multi-plates in EGM-2 medium with 10% FBS (final volume each well 100 μL). Add 100 μL of PBS or distilled water to the outside wells (not used for the experiment) to avoid evaporation of solutions.

4. Allow cell adherence in incubator at 37 °C for 4–5 h.

5. Prepare appropriate H₂S donors and inhibitors and keep on ice.

6. Remove media and treat cells with H₂S donors (1–100 μM) in the absence or presence of PAG (3 mM) or other metabolic inhibitors. Carry out the treatment in EBM-2 with low serum

concentration (0.1%) in a final volume of 100 μL. Repeat each experimental point three times.

7. Incubate the multi-plates at 37 °C and 5% CO_2 for 2–5 consecutive days with H_2S donors freshly added every 2 days.

8. Remove supernatants gently and fix the cells with methanol (100 μL; 4 °C, for 2 h).

9. Remove methanol and add Diff-Quik solution I (eosinophilic, orange dye); after 5 min, take the solution away and add Diff-Quik solution II (basophilic, blue dye) for other 5 min. Wash the multi-plate with water and let it to dry. Perform all these passages at RT.

10. Count cells randomly at 20× original magnification in five fields with the aid of an ocular grid (21 mm^2).

3.5 Cell Migration (Scratch Assay)

Migration of adherent cells can be evaluated following incubation with H_2S donors or other stimuli in the absence/presence of CSE or other metabolic inhibitors to assess the relevance of exogenous or endogenous H_2S in the promotion of cell migration [8, 9].

1. Harvest cells from Petri dish as described in Subheading 3.1.2, and plate in 24-well plates at 1×10^5 cells/well.

2. Incubate at 37 °C in a humidified incubator at 5% CO_2 in air. Allow cells to become 100% confluent.

3. Prepare appropriate H_2S donors and inhibitors and keep on ice.

4. Scrape cells in the central area of the well to create a wound of ±1 mm width using the sterile tip of a Gilson pipette.

5. Wash the surface once with PBS to remove detached cells.

6. Add the compounds previously prepared and the antimitotic ARA-C (2.5 μg/mL) to block cell proliferation ensuring that wound closure is due solely to migration.

7. Acquire duplicate images of the wound in each well under phase contrast microscope at 10× magnification from time 0 (to measure the initial wound width) for up to 18 h.

8. Measure area of wound using ImageJ software (see **Note 14**).

9. Results are expressed as percentage of wound area respect to time 0.

3.6 Cell Chemotaxis and Chemoinvasion

The Boyden Chamber procedure is used to evaluate cell chemotaxis and chemoinvasion [5]. The method is based on the passage of endothelial cells across porous filters against a concentration gradient of the migration effector. While in chemotaxis, the filter is coated with extracellular matrix (ECM) proteins, during chemoinvasion cells have to degrade a layer of ECM stratified on the filter. The Boyden Chamber consists of two parts separated by a polyvinylpyrrolidone (PVP)-free polycarbonate filter, 8 μm pore size, coated with type I collagen and fibronectin for chemotaxis and with 1% gelatin solution for chemoinvasion.

1. Filter coating: (a) for chemotaxis, incubate the filter with type I collagen (100 µg/mL) for 1 min at RT; dry under hood; incubate with bovine serum fibronectin (10 µg/mL) for 2–3 min at RT; wash with EBM medium to remove a specific binding. (b) For chemoinvasion, incubate the filter in 1 mL of 1% gelatin solution for 2–3 min at 37 °C; dry under hood.

2. Remove growth medium and apply trypsin/EDTA to cells as previously described (*see* Subheading 3.1.2 to remove cells from the Petri dish).

3. Count viable cells using a hemocytometer and adjust cell concentration to 2.5×10^5 cells/mL.

4. Prepare appropriate H$_2$S donors and inhibitors and keep on ice.

5. Put the cell suspension in sterile tubes (500 µL for each experimental point) for 30 min with PAG or vehicle at 37 °C.

6. Prepare chemoattractant solution (serum or angiogenic factors dissolved in EBM-2), and place them in the lower wells of Boyden Chamber (30 µL/well). Include in the experiment at least one experimental point that does not contain any chemoattractants. Do each experimental point in triplicate.

7. Cover the lower chamber with the pre-coated filter, and rapidly fix it with the specific support.

8. Add H$_2$S donor to cell suspension with or without inhibitor in sterile tubes, and then put 50 µL of treated cell suspension to each upper well.

9. Incubate the chamber at 37 °C in a humidified atmosphere of 5% CO$_2$ in air for 4 h for chemotaxis and 18 h for chemoinvasion.

10. Remove the filter and fix the cells in methanol overnight at 4 °C in a Petri dish.

11. Remove non-migrated cells on the upper surface of the filter with a cotton swab.

12. Stain migrated cells in the lower surface of the filter with Diff-Quik as described in Cell Proliferation section.

13. Cut the filter in four parts, and mount each of them on a histological slide covered with cover glass.

14. Count cells randomly at 40× original magnification in five fields/well with the aid of an ocular grid (21 mm^2).

15. Report data as migrated cells/well.

3.7 Gelatin Zymography

The production of degradative enzymes is evaluated in media from sub-confluent cell monolayers stimulated with H$_2$S donors in the absence/presence of metabolic inhibitors [9]. Cells secrete proteolytic enzymes which are loaded in polyacrylamide gels containing enzyme substrate as gelatin. After electrophoresis, the proteolytic enzymes are allowed to degrade gelatin. Finally the gel is stained with Coomassie Brilliant Blue dye to detect the proteolytic activity as white bands in a blue background. The procedure is divided into three phases: supernatant collection from cells, preparation of polyacrylamide gel containing gelatin, and detection of gelatinolytic activity.

3.7.1 Supernatant Collection

1. Remove growth medium and apply trypsin to cells as previously described (*see* Subheading 3.1.2 to remove cells from the Petri dish).

2. Count viable cells using a hemocytometer and adjust cell concentration to 5×10^3 cells/well.

3. Dispense a 100 μL volume of cells into each well of a 96-well plate with the exception of the outside wells. Make sure that each experimental point is repeated in triplicate. Add 100 μL of PBS or distilled water to the outside wells to maintain humidity.

4. After adherence, remove growth medium, and replace with basal medium without growth factors and with low serum (0.1%) for 16 h.

5. Prepare appropriate H_2S donors and inhibitors and keep on ice.

6. Remove supernatants from cells and add fresh basal medium without serum. Pretreat cells with inhibitors (30 min), and then add H_2S donors for 24 h (*see* **Note 15**).

7. Collect supernatants and immediately keep on ice or store at $-20\,^{\circ}C$ if assessed later.

8. Fix cell monolayers in methanol (2 h at 4 °C) and stain with Diff-Quik (as in Cell Proliferation section).

9. Count the total number of cells/well as described in Subheading 3.4.

3.7.2 Preparation of Polyacrylamide Gel Containing Gelatin and Gel Electrophoresis

1. Mix 2.5 mL of resolving buffer with 10% SDS, 2 mL of polyacrylamide, 1 mL of 1% gelatin solution, and 4.5 mL water in a 50 mL conical flask. Add 100 μL of APS and 10 μL of TEMED and cast gel within a 1.5 mm thickness gel cassette. Allow space for stacking gel and gently overlay with 20% methanol solution.

2. Allow gel polymerization and then remove methanol.

3. Prepare the stacking gel by mixing 1.25 mL of stacking gel buffer with 10% SDS, 0.56 mL of polyacrylamide, and 3 mL water in a 50 mL conical flask. Add 50 μL of APS and 5 μL of TEMED. Insert a 10-well gel comb immediately without introducing air bubbles.

4. Place the 8% polyacrylamide gel containing gelatin in a Mini-protean tetra system Bio-rad as described in the manufacturer's instructions.

5. Fill tank with cold SDS-PAGE running buffer to recommended level and put the Mini-protean tetra system Bio-rad on ice.

6. Mix 50 μL of supernatants (prepared in Subheading 3.7.1) with 12.5 μL of 4× Laemmli sample buffer without heating the samples.

7. Add samples to the wells in the gel, including molecular weight standard.

8. Connect electrophoresis equipment to power pack, and run at 80 V for 10 min until the samples have entered in the separating gel and then continue at 180 V till the dye front reaches the bottom of the gel.

9. Open the gel plates with the use of a spatula and rinse the gel with washing solution.

3.7.3 Detection of Gelatinolytic Activity

1. Wash gel twice with gentle agitation for 15 min with 2.5% Triton X-100 to remove SDS.

2. Add developing solution and incubate 48 h at 37 °C in an incubator.

3. Stain with Coomassie Brilliant Blue R-250 solution for 1 h with gentle agitation.

4. Destain using destaining solution replaced every 15 min with gentle agitation.

5. Place the gel on Chemidoc and acquire an image of the gel using a digital camera.

6. Normalize the optical density values of the white bands to the number of cell counted/well.

3.8 Matrigel Plug Assay

A method to analyze sprouting and network formation by endothelial cells is based on the use of Matrigel [8, 9]. Matrigel is a heterogeneous mixture of matrix proteins localized in the basal membrane such as collagen type IV, laminin, and heparan sulfate and is resistant to digestion by endogenous enzymes.

1. Place Matrigel on ice at 4 °C, and leave overnight to thaw completely, taking care not to let it warm at RT.

2. Cut the top of any tips that will be used for Matrigel manipulation.

3. Mix Matrigel to homogeneity by gentle pipetting.

4. Add 250 μL of Matrigel to each well of a 24-well plate.

5. Place the plate at 37 °C for at least 30 min to leave the Matrigel to solidify.

6. Harvest endothelial cells silenced for CSE (in 6-well plates, as described in Subheading 3.2) or from Petri dish as described in Subheading 3.1.2.

7. Prepare appropriate H_2S donors and inhibitors and keep on ice.

8. Plate 7×10^4 silenced cells or treated with H_2S donors and inhibitors onto the thin layer of basement membrane matrix Matrigel.

9. Include 0.1% FBS as negative control and 10% FBS or angiogenic factors as positive control.

10. Incubate plate for ~16 h at 37 °C in a humidified incubator at 5% CO_2 in air.

11. Capture images using a brightfield microscope at 4× and 10× magnification in ten fields.

12. In order to quantify the networks, count the number of nodes, evaluating the number of branch points.

13. Report results as branch points/well.

4   Notes

1. It is necessary to prepare NaHS fresh solutions each time. NaHS is the most popular H_2S donor, but it is possible to use other compounds as GYY4137 slow releasing H_2S donor [15], AP39, mitochondria-targeted H_2S donor [16], and active metabolite of drug containing SH moiety as zofenoprilat [8, 9].

2. Once solubilized, it is possible to store aliquots of PAG stock solution at −20 °C for 6 months.

3. Human endothelial cells are commercially available and can be obtained from large vessels or microvasculature, from specialized vascular areas such as umbilical vein of newborn and blood vessels of specific organs or from solid tumors. Moreover, it is possible to have cells from healthy subject or, for example, from diabetic patients. Endothelial cells from umbilical vein are the most commonly used and usually derived from pooled donors. As angiogenesis originated from microvasculature, it is suitable to perform key experiments in such cells [17].

4. It is important to use primary endothelial cells at low passage as cells became senescent and lose typical endothelial phenotype.

5. The gelatin solution is prepared as follows: the powder of bovine gelatin is weighted (5 g) and dissolved in 500 mL of

Milli-Q water. To guarantee sterility, the solution is autoclaved and filtered (0.2 μm filter) in sterile 50 mL tubes. The stock solution can be maintained at 4 °C for 1 month.

6. Ambion Silencer Negative Control is a nontargeting gene with limited sequence similarity to known genes. This is validated for use in human, mouse, and rat cells. It is functionally proven to have minimal effects on cell proliferation and viability. This is HPLC purified, duplexed, and ready to use as the negative control siRNAs are essential for determining transfection efficiency and to control for the effects of siRNA delivery.

7. The fluorescent dye WSP-1 (Cayman Chemical, Ann Arbor, MI, USA) was used to assess H$_2$S levels directly released in vitro or produced by cells, being released by the cells in the culture medium or in the cytoplasm [9]. Through a reaction-based fluorescent turn-on strategy, WSP-1 selectively and rapidly reacts with H$_2$S to generate benzodithiolone and a fluorophore with excitation and emission maxima of 465 and 515 nm, respectively [18].

8. To be prepared fresh each time.

9. We find that storing TEMED at 4 °C reduces its pungent smell.

10. Store concentrated Matrigel at −20 °C. Place the gel on ice in a 4 °C refrigerator to allow slow thawing overnight before use.

11. Cut the top of the tips to adequately pipette Matrigel in 24-well tissue culture plates.

12. Trypsin-EDTA treatment is the most common method for cell detachment. Prolonged exposure to trypsin is deleterious for cells. To avoid this, it is important to inactivate the enzyme with serum, e.g., using this proportion: 700 μL of serum for each 2 mL of trypsin-EDTA.

13. Prepare WSP-1 diluting 2 mg of WSP-1 in 1 mL of sterile DMSO.

14. Open the files with images of time 0 and time 18 h for each experimental points under ImageJ software. Using freehand selection command, draw the outline of the wound, and then measure the area (Ctrl + M). Report each area values on Excel file. The area value of time 0 is taken as 100% to calculate the percentage change in wound width of the area calculated for the image corresponding to 18 h. The lower is the percentage, the higher is cell migration. Repeat the procedure for each image set.

15. HT1080 cells can be used as gold standard for the metalloproteinases MMP-2 and MMP-9 activity,

Acknowledgments

Part of this work was funded by Italian Space Agency (project "Tissue Repair in Microgravity" ASI N. 2013-090-R.O).

References

1. Chung AS, Ferrara N (2011) Developmental and pathological angiogenesis. Ann Rev Cell Dev Biol 27:563–584

2. Welti J, Loges S, Dimmeler S, Carmeliet P (2013) Recent molecular discoveries in angiogenesis and antiangiogenic therapies in cancer. J Clin Invest 123(8):3190–3200

3. Ferrara N, Adamis AP (2016) Ten years of anti-vascular endothelial growth factor therapy. Nat Rev Drug Discov 15(6):385–403. https://doi.org/10.1038/nrd.2015.17

4. Bachetti T, Morbidelli L (2000) Endothelial cells in culture: a model for studying vascular functions. Pharmacol Res 42:9–19

5. Ziche M, Morbidelli L, Masini E et al (1994) Nitric oxide mediates angiogenesis in vivo and endothelial cell growth and migration in vitro promoted by substance P. J Clin Invest 94:2036–2044

6. Morbidelli L, Chang C-H, Douglas JG et al (1996) Nitric oxide mediates mitogenic effect of VEGF on coronary venular endothelium. Am J Phys 39(1):H411–H4156

7. Morbidelli L, Pyriochou A, Filippi S et al (2010) The soluble guanylyl cyclase inhibitor NS-2028 reduces vascular endothelial growth factor-induced angiogenesis and permeability. Am J Physiol Regul Integr Comp Physiol 298(3):R824–R832

8. Terzuoli E, Monti M, Vellecco V et al (2015) Characterization of zofenoprilat as an inducer of functional angiogenesis through increased H_2S availability. Br J Pharmacol 172(12):2961–2973. https://doi.org/10.1111/bph.13101

9. Monti M, Terzuoli E, Ziche M, Morbidelli L (2016) H_2S dependent and independent anti-inflammatory activity of zofenoprilat in cells of the vascular wall. Pharmacol Res 113 (Pt A):426–437. https://doi.org/10.1016/j.phrs.2016.09.017

10. Kimura H (2011) Hydrogen sulfide: its production, release and functions. Amino Acids 41 (1):113–121. https://doi.org/10.1007/s00726-010-0510-x

11. Wallace JL, Wang R (2015) Hydrogen sulphide-based therapeutics: exploiting an unique but ubiquitous gasotransmitter. Nat Rev 14:329–345

12. Yang G, Wu L, Jiang B et al (2008) H_2S as a physiologic vasorelaxant: hypertension in mice with deletion of cystathionine gamma-lyase. Science 322(5901):587–590. https://doi.org/10.1126/science.1162667

13. Yang G, Wang R (2015) H_2S and blood vessels: an overview. Handb Exp Pharmacol 230:85–110

14. Katsouda A, Bibli SI, Pyriochou A et al (2016) Regulation and role of endogenously produced hydrogen sulfide in angiogenesis. Pharmacol Res 113(Pt A):175–185. https://doi.org/10.1016/j.phrs.2016.08.026

15. Li L, Whiteman M, Guan YY, Neo KL et al (2008) Characterization of a novel, water-soluble hydrogen sulfide-releasing molecule (GYY4137): new insights into the biology of hydrogen sulfide. Circulation 117(18):2351–2360. https://doi.org/10.1161/CIRCULATIONAHA.107.753467

16. Szczesny B, Módis K, Yanagi K et al (2014) AP39, a novel mitochondria-targeted hydrogen sulfide donor, stimulates cellular bioenergetics, exerts cytoprotective effects and protects against the loss of mitochondrial DNA integrity in oxidatively stressed endothelial cells in vitro. Nitric Oxide 41:120–130. https://doi.org/10.1016/j.niox.2014.04.008

17. Monti M, Solito R, Puccetti L et al (2014) Protective effects of novel metal-nonoates on the cellular components of the vascular system. J Pharmacol Exp Ther 351(3):500–509. https://doi.org/10.1124/jpet.114.218404

18. Liu C, Pan J, Li S et al (2011) Capture and visualization of hydrogen sulfide via a fluorescent probe. Angew Chem Int Ed Engl 50(44):10327–10329

Chapter 12

Central Administration of H$_2$S Donors for Studying Cardiovascular Effects of H$_2$S in Rats

Marcin Ufnal and Artur Nowinski

Abstract

Increasing evidence suggests that hydrogen sulfide (H$_2$S) is involved in brain mechanisms regulating the functions of the circulatory system. This appears to be mediated by cardiovascular centers located in the central nervous system. This chapter describes techniques of acute and chronic infusions into the brain cardiovascular centers in rats. Rats may be implanted either acutely or chronically with a cannula inserted into a selected cardiovascular center according to the stereotaxic coordinates. The cannula allows for the administration of the investigated compounds into a selected cardiovascular center.

Key words Hydrogen sulfide, Central infusions, Intracerebroventricular, Circumventricular organs, Paraventricular nucleus, Rostral ventrolateral medulla, Caudal ventrolateral medulla

1 Introduction

H$_2$S plays a role as a mediator in the brain cardiovascular centers that control the circulatory system by modulating the activity of preganglionic neurons of the autonomic nervous system [1]. Major brain cardiovascular centers include the circumventricular organs, the paraventricular nucleus, the rostral ventrolateral medulla, the caudal ventrolateral medulla, and others [2]. To evaluate the involvement of H$_2$S in the nervous control of the circulatory system, infusions of H$_2$S donors or inhibitors of H$_2$S enzymatic production into cerebroventricular system of the brain or into a given cardiovascular center may be performed. Intracerebroventricularly infused compounds target the circumventricular organs and may penetrate to deeper brain structures, depending on chemical properties of the investigated compounds [3]. Infusions into a specific brain region allow to evaluate the effects of administered compounds on a selected brain cardiovascular center. For assessing the role of blood-borne H$_2$S donors or inhibitors of H$_2$S enzymatic production, the first method offers closer to real-life conditions. This is because under most physiological conditions, the

Jerzy Bełtowski (ed.), *Vascular Effects of Hydrogen Sulfide: Methods and Protocols*, Methods in Molecular Biology, vol. 2007,
https://doi.org/10.1007/978-1-4939-9528-8_12, © Springer Science+Business Media, LLC, part of Springer Nature 2019

compounds circulating in the blood and/or in the cerebrospinal fluid first target the circumventricular organs which lack the blood-brain barrier. Intracerebroventricular infusions are performed by means of a cannula inserted into the lateral cerebral ventricle. The techniques of intracerebroventricular infusions and selective infusions into a specific cardiovascular center differ in the selection of stereotactic coordinates before the insertion of the infusing cannula. During the brain infusions, hemodynamic and electrocardiographic parameters may be recorded (the implantation of arterial catheters and ECG electrodes has been described in Chapter 13).

2 Materials

1. H_2S donor.
2. Guide cannula, no larger than ID 0.7 mm × OD 0.9 mm (acute infusions).
3. Infusion cannula, no larger than ID 0.5 mm × OD 0.6 mm (acute infusions).
4. Metal stylet (acute infusions, *see* **Note 1**).
5. Dental cement, cyanoacrylate adhesive.
6. Brain infusion kit (e.g., the one manufactured by Alzet, Cupertino, CA, USA).
7. Syringe infusion pump (acute infusions).
8. Microsyringes, e.g., Hamilton syringes (acute infusions).
9. Osmotic or mechanic subcutaneous minipumps (chronic infusions).
10. Nylon skin sutures 5-0 or 4-0.
11. Polyethylene tubings to connect a microsyringe with the infusion cannula (ID 0.5 mm × OD 0.8 mm—ID matching infusion cannula OD).
12. Sterile needles, various size 0.5–1.2 mm.

3 Methods

3.1 Stereotaxic Surgery

3.1.1 Pre-surgical Preparation

All procedures should be performed in a possibly aseptic way. The procedures differ slightly, depending on the choice of a guide cannula, used for acute infusions (Subheading "Acute Infusions"); the brain infusion kit, used for both acute and chronic studies; and the brain infusion kit with a minipump, used for chronic delivery (Subheading "Chronic Infusions").

1. Clean and disinfect the workplace and stereotaxic apparatus, e.g., with 70% water ethanol solution.

2. Prepare all the needed instruments (which should be sterilized) and materials. Use heating pads to prevent a drop in body temperature.

3. Weigh the animal to prepare an adequate dose of anesthetic (e.g., 60 mg/kg ketamine and 10 mg/kg xylazine injected intraperitoneally). Wait until the animal is unconscious and proceed to the next step (*see* **Note 2**).

4. Shave the rat's head with an electric razor beginning from ears and finishing just behind the eyes. Do it as thoroughly as possible.

5. Check if the rat is completely anesthetized (by toe pinch or using other methods like tail reflex). If so, proceed to the next point. Otherwise, some extra anesthetic may be needed.

6. Place the rat in the stereotaxic apparatus. Ensure that ear bars are inserted correctly and that the animal's head is stable, and then fix the nose bar. Make sure the head is level.

7. If you are going to use the brain infusion kit, make sure to use enough spacers to get to the desired depth. Prepare the infusion pump according to the manufacturer's instructions.

3.1.2 Cannula Implantation

1. Clean and disinfect the shaved part of the skin with 70% ethanol solution (or other agents like chlorhexidine or povidone/iodine solutions).

2. Make a longitudinal incision about 2–2.5 cm long in the midline of the rat's head.

3. Hold the skin edges to keep the incision open. Completely clean the conjunctive tissues covering the bone. Use cotton swabs to dry off the skull surface and ensure hemostasis.

4. Place the cannula tip over bregma touching the skull surface and record the coordinates (if your equipment allows it, reset your coordinates, so that bregma is "0" point). Then add or subtract coordinates, preferably taken from a stereotaxic atlas, to place the cannula in the appropriate position to access the desired brain structures (*see* **Note 3**). Mark the position with a sterile pencil.

5. Carefully drill a hole in the skull using a sterile drill bit (alternatively a needle) at the pencil mark location. The hole should be at least the same diameter as the cannula or slightly larger. Puncture meninges with a sterile needle.

Acute Infusions

6. Clean the cannula with the ethanol solution and rinse with saline. Place the cannula carefully until reaching the desired dorsoventral coordinate. To anchor the cannula, you can additionally use two metal screws.

7. Prepare dental cement, and apply it to cover the skull and the base of the cannula (screws also, if used) (*see* **Note 4**).

8. When the cement is completely dry (usually it takes a few minutes), detach the cannula from the apparatus. Insert a sterile stylet into the cannula canal to prevent obstruction. Remove the rat from the apparatus for easier manipulations during stitching.

9. Using nylon skin sutures, stitch the skin in at least two places, one anterior and one posterior to the cannula.

Chronic Infusions

Skip **steps 6–9** and proceed to **step 10**.

10. Cover the bottom surface of the spacers with cyanoacrylate adhesive, and then carefully but firmly place the infusion kit in the desired position.

11. Create a subcutaneous pouch caudally from the posterior part of the wound for the minipump implantation.

12. Implant the minipump subcutaneously (using the pouch created just before) after connecting it to the infusion kit via a polyurethane catheter.

13. Using the nylon skin sutures, stitch the skin in at least two places (*see* **Note 5**).

3.1.3 Postimplantation Period

1. After the surgery, inject 25,000–30,000 units of benzathine penicillin intramuscularly and 1–2 ml of warm saline subcutaneously.

2. Observe the animal carefully during the first hours after surgery, especially in the first 30 min, until the rat is fully conscious.

3. Use antibiotic ointment on the wound.

4. Monitor the animal after the surgery for a few days.

5. If the rat does not look healthy, do not use it in the experiment.

Additional reading for stereotaxic surgery [4, 5].

3.2 Acute Infusions

1. Prepare the H_2S donor.

3.2.1 Preparation

2. Prepare a microsyringe and an infusion pump.

3. Prepare an infusion cannula and connect it to a flexible polyethylene tubing.

4. Fill the microsyringe with H_2S donor, and connect it to the infusion cannula via the tubing.

5. Fill the whole set with the solution from the microsyringe, so that no air is visible.

6. Put the syringe into the infusion pump; set the pump to the desired flow rate.

3.2.2 Infusion

1. Remove the metal stylet from the previously implanted guide cannula.

2. Insert the infusion cannula into the guide cannula.

3. Begin the infusion (*see* **Note 6**).

4 Notes

1. It may be done using any sterile cannula of a size that matches the internal size of the guide cannula, or a manufactured dummy cannula may be used.

2. Other concentrations may be used, depending on the research center, usually between 50–100 mg/kg ketamine and 5–10 mg/kg xylazine [6, 7]. Other anesthetic agents or mixtures may be used. Gas anesthesia with isoflurane can also be used.

3. The commonly used coordinates for lateral ventricle are −0.12 mm AP (anteroposterior), 1.6 mm ML (mediolateral), and 4.3 mm DV (dorsoventral) or −1.0 mm AP, 2.4 mm ML, and 4.0 mm DV [8, 9]; for paraventricular nucleus, the example coordinates are −1.6 mm AP, 0.2 mm ML, and 7.2 mm DV [10]; coordinates can be calculated using the stereotaxic atlas, and because of the small size of neurovascular centers, the coordinates strongly depend on animals' weight and strain, so the accuracy should be histologically confirmed in each experiment. A useful online tool can be found at http://labs.gaidi.ca/rat-brain-atlas/ [11].

4. It is the good moment to remove the excess cement and to shape it.

5. Be careful not to puncture the connecting tube (it is right beneath the skin).

6. Maximal infusion rate for intracerebroventricular infusions is about 40 μl/h and no more than 0.1–0.2 μl/min for infusions into a specific cardiovascular center [6, 12, 13].

References

1. Ufnal M, Sikora M (2011) The role of brain gaseous transmitters in the regulation of the circulatory system. Curr Pharm Biotechnol 12 (9):1322–1333

2. Dampney RA, Coleman MJ, Fontes MA et al (2002) Central mechanisms underlying short- and long-term regulation of the cardiovascular system. Clin Exp Pharmacol Physiol 29 (4):261–268

3. Ufnal M, Skrzypecki J (2014) Blood borne hormones in a cross-talk between peripheral and brain mechanisms regulating blood pressure, the role of circumventricular organs. Neuropeptides 48(2):65–73. https://doi.org/10.1016/j.npep.2014.01.003

4. Fornari RV, Wichmann R, Atsak P et al (2012) Rodent stereotaxic surgery and animal welfare outcome improvements for behavioral

neuroscience. J Vis Exp (59):e3528. https://doi.org/10.3791/3528

5. Geiger BM, Frank LE, Caldera-Siu AD et al (2008) Survivable stereotaxic surgery in rodents. J Vis Exp (20). https://doi.org/10.3791/880

6. Ufnal M, Sikora M, Dudek M (2008) Exogenous hydrogen sulfide produces hemodynamic effects by triggering central neuroregulatory mechanisms. Acta Neurobiol Exp (Wars) 68 (3):382–388

7. Struck MB, Andrutis KA, Ramirez HE et al (2011) Effect of a short-term fast on ketamine-xylazine anesthesia in rats. J Am Assoc Lab Anim Sci 50(3):344–348

8. Paxinos G, Watson C (2005) The rat brain in stereotaxic coordinates. Academic, San Diego

9. Lourbopoulos A, Grigoriadis N, Karacostas D et al (2010) Predictable ventricular shift after focal cerebral ischaemia in rats: practical considerations for intraventricular therapeutic interventions. Lab Anim 44(2):71–78. https://doi.org/10.1258/la.2009.009043

10. McMahon LR, Wellman PJ (1998) PVN infusion of GLP-1-(7-36) amide suppresses feeding but does not induce aversion or alter locomotion in rats. Am J Phys 274(1 Pt 2): R23–R29

11. Matt G (2012) Rat brain atlas. http://labs.gaidi.ca/rat-brain-atlas/. Accessed 19 Jan 2017

12. Peterson SL (1998) Drug microinjection in discrete brain regions. http://kopfinstruments.com/app/uploads/2015/04/Carrier50.pdf. Accessed 24 Jan 2017

13. Carvey PM, Maag TJ, Lin D (1994) Injection of biologically active substances into the brain. In: Flanagan TRJ, Emerich DF, Winn SR (eds) Providing pharmacological access to the brain: alternate approaches. Methods in neurosciences, vol 21. Academic, San Diego, pp 214–236

Chapter 13

Colonic Delivery of H$_2$S Donors for Studying Cardiovascular Effects of H$_2$S in Rats

Marcin Ufnal and Tomasz Hutsch

Abstract

There is evidence that H$_2$S produced in the colon may contribute to the control of the circulatory system. This chapter describes a technique of cardiovascular measurements in anesthetized rats subjected to intracolonic administration of H$_2$S donor. The intracolonic administration is performed via polyurethane catheter inserted per rectum into the colon.

Key words Hydrogen sulfide, Colon, i.c., Cardiovascular diseases, Gut, Bacteria

1 Introduction

So far, cardiovascular research on H$_2$S has focused on the effects of H$_2$S enzymatically produced by mammalian tissues. However, there is some evidence that H$_2$S released by bacteria in the gut may also contribute to the circulatory system homeostasis [1, 2]. Interestingly, sulfate-reducing bacteria, such as *Desulfovibrio niger*, are ubiquitous in mammalian colon, producing substantial amounts of H$_2$S. It is worth noting that while the tissue enzymes can produce H$_2$S in nanomolar range, gut bacterial flora may produce H$_2$S in millimolar range [2]. Intracolonic infusions of H$_2$S donors or H$_2$S-deactivating compounds may be an interesting research tool in studying cardiovascular effects of colon-derived H$_2$S. Intracolonic administration may be performed in anesthetized rats implanted with arterial catheter and electrodes for hemodynamic and electrocardiographic recordings, respectively. The intracolonic administration is performed via a flexible polyurethane tube inserted into the colon 6–10 cm from the external anal sphincter.

Jerzy Bełtowski (ed.), *Vascular Effects of Hydrogen Sulfide: Methods and Protocols*, Methods in Molecular Biology, vol. 2007, https://doi.org/10.1007/978-1-4939-9528-8_13, © Springer Science+Business Media, LLC, part of Springer Nature 2019

2   Materials

2.1   Anesthetics and Drugs

1. Urethane.
2. Ketamine.
3. Xylazine.
4. Heparin for i.v. infusions.
5. 2% xylocaine.
6. 0.9% NaCl.
7. Povidone-iodine (PVP-I).

2.2   Catheters

1. Arterial catheters: flexible polyurethane catheters of various sizes (depending on animal size), e.g., ID: 0.635 mm × OD: 1.02 mm or ID: 0.305 mm × OD: 0.635 mm.
2. Colonic catheter: flexible polyethylene catheter OD: 2.1 mm.
3. Lubricant, e.g., glycerin or petrolatum.

2.3   Surgical Instruments

1. Surgical drapes.
2. Surgical silk sutures 3-0.
3. Microsurgical scissors.
4. Tweezers with curved and blunt ends.
5. Injection needles 0.9 × 4.0 mm.
6. Other basic surgical instruments.

3   Methods

3.1   Implantation of Arterial Catheters

1. To perform direct measurements of blood pressure, insert a polyethylene catheter into the abdominal aorta via the femoral artery. Depending on the size of an animal (size of arteries), use catheters of various sizes, e.g., a thin catheter (OD: 0.635 mm × 1.02 mm) or a thick catheter (OD: 0.305 mm × 0.63 5 mm).
2. Before inserting the catheter into the artery, fill the catheter with heparin (heparin solution in 0.9% NaCl, 1:5 ratio).
3. Anesthetize the rat with intraperitoneal injection of urethane 1.5 g/kg of body weight (BW) (terminal studies) or other anesthetics, e.g., xylocaine 5–10 mg/kg BW and ketamine 40–80 mg/kg BW. Solution of xylazine/ketamine can be prepared in one syringe (see **Note 1**).
4. Shave fur in the groin and disinfect the skin with povidone-iodine (PVP-I); cover the groin area with surgical drapes.

5. Cut the skin longitudinally for the length of about 2.0 cm in the place where the pulse of the femoral artery is palpable.

6. Dissect the muscles and fascia to visualize the neurovascular bunch.

7. Use tweezers with blunt ends to dissect the femoral artery from the neurovascular bunch: first nerves, then the femoral artery, and then the vein (*see* **Note 2**).

8. Put two ligatures on the femoral artery. With the distal ligature, tie the surgical knot on the distal end of the femoral artery. Catch the ends of the proximal ligature with a needle holder. Carefully pull the ligature ends with the holder upward to close the proximal part of the artery. Do not apply too much force and do not tie the knot! Secure the holder on the surgical table above the animal's head keeping the artery closed.

9. Apply a drop of xylocaine on the artery. The artery will dilate.

10. Use microsurgical scissors to make a small incision on the femoral artery, between the knot and proximal ligature. Insert the catheter using tweezers. Alternatively you may use an injection needle with the curved end to puncture the artery and use the bended tip of the needle as a guide for the catheter.

11. Insert the catheter for 2–4.0 cm (depending on the size of an animal) to place the proximal tip of the catheter in the abdominal aorta, approx. 2.0 cm below the branching of the renal arteries (*see* **Notes 3–6**).

12. Secure the catheter in the femoral artery with two single surgical knots. Flash the line immediately after insertion of the catheter with heparinized saline (100 units/mL) to prevent formation of a blood clot. Close the surgical wound with two layers of single stitches (*see* **Notes 7 and 8**).

3.2 Electrocardiogram Recordings

1. Use standard needle electrodes for electrocardiogram recordings.

2. Insert the electrodes under the skin of left and right forepaws and the tail. Additionally, unipolar leads can be positioned anteriorly at the midstream [3].

3.3 The Catheterization of the Colon

1. Before inserting the catheter into the colon, check the following: anesthesia, anal region, and the content of the stools in the rectum. If the stools are present, empty the rectum by massaging the rectal area.

2. The length of the catheter should be 12.0–15.0 cm. Mark the catheter to indicate the part which will be inserted into the colon (6–10 cm). This is necessary to control the proper depth of the catheter insertion into the colon.

3. Put a lubricant (e.g., glycerin) along all the catheter part that will be inserted into the colon. The anus should also be moistened with the lubricant.

4. Insert the catheter 6–10 cm through the external anal sphincter. Insert the catheter slowly—make forward-backward and circular movements (*see* **Note 9**).

5. Use a sticking plaster to fix the distal part of the catheter to the rat's tail. This will secure proper placement of the catheter throughout the experiment (*see* **Note 10**).

3.4 Measurements

1. Perform the experiments evaluating the cardiovascular effects of intracolonic administration of an H_2S donor in animals under general anesthesia.

2. Start the experimental measurements no earlier than 15 min after inserting the intracolonic catheter and connecting the arterial catheter to the recording system.

4   Notes

1. To facilitate absorption of anesthetics, the volume of drugs can be increased with 0.9% saline (to 1 mL), and the abdominal massage may be performed after administration of the drugs.

2. The femoral artery and vein often have several tiny branches which are located between the muscles. You need to be careful since during the dissection of the neurovascular bundle, those vessels may easily be damaged producing bleeding.

3. If you can't see the incision site on the artery, you can lose the proximal ligature for a second. If the artery is punctured, a spot of blood will appear on the artery.

4. If the artery is contracted, put a drop of lidocaine and wait until the artery dilates. It may also be helpful to inject a small drop of xylocaine into the lumen of the artery.

5. If the catheter cannot be inserted through the first incision site, make a second incision above the first incision (more proximal).

6. The intravascular part of the catheter must not be too long. If the tip of the catheter riches abdominal aorta at the level of renal arteries, it may reduce the renal blood flow. This will result in the activation of the renin-angiotensin system and hemodynamic and biochemical disturbances.

7. After inserting the catheter into the artery, the catheter patency must be checked. Blood in the catheter should recede spontaneously. Otherwise, you may expect:

   (a) A blood clot in the catheter.

   (b) The catheter being bent inside the vessel.

   (c) Artery wall perforation.

   (d) Abnormal anatomical structure of the vessel.

   (e) Too tight surgical knot which closes the lumen of the artery.

   You may try to:

   (a) Flush the catheter with heparin solution.

   (b) Gently pull the catheter 1–2 mm from the artery.

   (c) Remove the surgical knots, and tie a new one.

   (d) Pull the catheter out and reinsert, or replace with a new catheter.

8. Air bubbles in the catheter will disturb the measurement. Backflow of the blood in the catheter connected to the recording systems usually indicates the leakage from the catheter system.

9. Do not use force while inserting a catheter into the colon because perforation of the intestine is possible. If you can't advance the catheter deep enough you may try to:

   (a) Inject about 0.5 mL of saline.

   (b) Leave the catheter in the colon for 5–10 min and try again.

10. It happens that the catheter bends in the rectum. Therefore, while inserting the catheter, keep on checking the location of the catheter by abdominal palpation.

References

1. Tomasova L, Dobrowolski L, Jurkowska H, Wrobel M, Huc T, Ondrias K, Ostaszewski R, Ufnal M (2016) Intracolonic hydrogen sulfide lowers blood pressure in rats. Nitric Oxide 60:50–58

2. Tomasova L, Konopelski P, Ufnal M (2016) Gut bacteria and hydrogen sulfide: the new old players in circulatory system homeostasis. Molecules 21(11)

3. Konopelski P, Ufnal M (2016) Electrocardiography in rats: a comparison to human. Physiol Res 65(5):717–725

Chapter 14

Measurements for Sulfide-Mediated Inhibition of Myeloperoxidase Activity

Dorottya Garai, Zoltán Pálinkás, József Balla, Anthony J. Kettle, and Péter Nagy

Abstract

Oxidative stress-alleviating and inflammation-mediatory functions of hydrogen sulfide were reported to be key features of its biological actions. However, the underlying molecular mechanisms of these biological observations are not fully understood. In conditions where sulfide was proposed to be protective against oxidative stress- or inflammation-induced tissue damage (e.g., reperfusion injury, atherosclerosis, vascular inflammation), the reactive oxidant-producing function of a key neutrophil enzyme, myeloperoxidase, was reported to be a protagonist on the detrimental side. We recently described favorable interactions between sulfide and myeloperoxidase and proposed that the potent inhibition of myeloperoxidase activities could contribute to sulfide's beneficial functions in a number of cardiovascular pathologies. Our chapter is dedicated to aid future studies and drug development endeavors in this area by providing methodological guidance on how to assess the inhibitory potential of sulfide on myeloperoxidase enzymatic activities in isolated protein systems, in neutrophil homogenates, and in live neutrophil preparations.

Key words Myeloperoxidase, Hydrogen sulfide, Peroxidase activity, Halogenation activity, Neutrophil function

Abbreviations

DMF	N,N-Dimethylformamide
DMSO	Dimethyl sulfoxide
DPBS	Dulbecco's phosphate buffered saline
DTNB	5,5-Dithio-bis-(2-nitrobenzoic acid)
DTPA	Diethylenetriaminepentaacetic acid
EDTA	Ethylenediaminetetraacetic acid
HBSS	Hank's Balanced Salt Solution
HRP	Horseradish peroxidase
HTAB	Hexadecyltrimethylammonium bromide
MPO	Myeloperoxidase enzyme
NOX2	Nicotinamide adenine dinucleotide phosphate (NADPH) oxidase 2 enzyme-complex
PBS	Phosphate-buffered saline
PMA	Phorbol 12-myristate 13-acetate

Jerzy Bełtowski (ed.), *Vascular Effects of Hydrogen Sulfide: Methods and Protocols*, Methods in Molecular Biology, vol. 2007,
https://doi.org/10.1007/978-1-4939-9528-8_14, © Springer Science+Business Media, LLC, part of Springer Nature 2019

RFU   Relative fluorescence unit
ROS   Reactive oxygen species
SOD   Superoxide dismutase
TMB   3,3′,5,5′-Tetramethylbenzidine
UV-Vis   Ultraviolet-visible

1   Introduction

Hydrogen sulfide has been recognized as an important bioactive molecule due to its role as a mediator of a plethora of physiological processes [1–3]. Some of the diverse biological properties of sulfide were associated with the production/scavenging of reactive oxygen species (ROS) [4–13]. These functions were found to have an emerging significance in inflammatory processes, particularly in cardiovascular disorders [14–16]. We argue that due to low endogenous free sulfide levels [17], the previously observed oxidative stress-alleviating effects of H_2S are unlikely to be associated with direct scavenging of ROS [18]. On the other hand, we proposed that inhibition of ROS production via interactions of sulfide with heme protein centers is a viable model [11]. In our previous study, we have shown that sulfide is a very effective inhibitor of the neutrophil granulocyte enzyme, myeloperoxidase (MPO) [19]. The primary function of MPO during phagocytosis is the destruction of pathogenic microorganisms by producing ROS inside the phagosomal space [20]. However, extracellular production of these oxidants has been associated with inflammation-induced tissue damage in various organs [21, 22]. In particular, MPO has a pro-inflammatory role in several cardiovascular diseases, and hydrogen sulfide was reported to be a protective agent in most of these conditions, for example, in reperfusion injury [16, 23], rheumatoid arthritis [15, 24], or atherosclerosis [14, 23, 25]. Therefore, we suggested that the interactions of sulfide with different enzyme forms of MPO could have important biological implications in particular in inflammatory diseases and vascular pathologies [19]. The aim of this chapter is to give detailed protocols on how to measure the effects of sulfide on the activities of MPO in isolated protein systems, in neutrophil homogenates, and in activated, live neutrophil samples.

The halogenation cycle of MPO consists of two major steps: In the first step, hydrogen peroxide reacts with native ferric MPO to form Compound I which is an oxyferryl species (Fe(IV)=O), containing a porphyrin π-cation radical. This enzyme form then can be reduced by halogenides in a subsequent two-electron redox reaction to produce hypohalous or pseudohalous acids and convert MPO back to its native ferric enzyme form to close the halogenation cycle (*see* Scheme 1) [26–28].

native enzyme

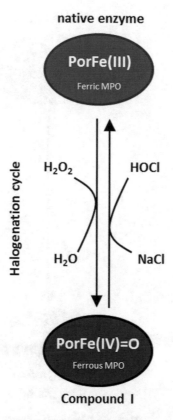

Scheme 1 Halogenation cycle of myeloperoxidase. In the first step of the halogenation cycle, hydrogen peroxide reacts with the native enzyme form to produce Compound I, which is an oxyferryl species with a porphyrin π-cation radical. Compound I oxidizes halogenides in a two-electron reaction that gives hypo(pseudo)halous acids and brings MPO back to the ferric enzyme form to close the chlorination cycle

In the peroxidase cycle of MPO H_2O_2 converts native MPO to Compound I (this first step is similar to the first step of the halogenation cycle). Compound I can then also react in one-electron redox reactions with various biological substrates including tyrosine to produce radical species with a concomitant formation of the Compound II enzyme form (a Fe(IV)-OH derivative) [29]. Compound II represents the resting state of the enzyme under physiological conditions, because in many cases the rate-determining step is its subsequent one-electron reduction to the ferric state to close the peroxidase cycle (*see* Scheme 2). In biological samples both the halogenation and peroxidase cycles operate, and their relative contributions mostly depend on substrate concentrations.

This chapter is dedicated to provide detailed protocols on how to measure the inhibitory actions of sulfide on the enzymatic activities of MPO. Because sulfide can react with various products and intermediate species that are generated during the reactions of

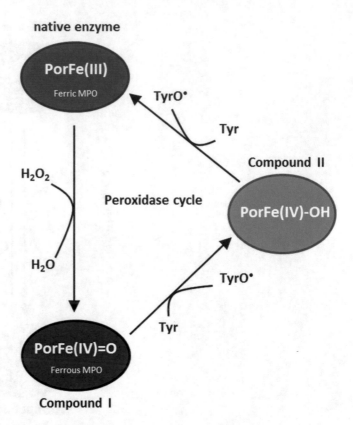

native enzyme

PorFe(III)
Ferric MPO

TyrO•

Tyr

H₂O₂

H₂O

Peroxidase cycle

Compound II

PorFe(IV)-OH

TyrO•

Tyr

PorFe(IV)=O
Ferrous MPO

Compound I

Scheme 2 Peroxidase cycle of myeloperoxidase. During the peroxidase cycle of MPO, Compound I (which is generated in the $2e^-$ oxidation of the native enzyme by H_2O_2) is reduced back to the ferric enzyme form by various substrates (e.g., tyrosine) in two consecutive one-electron reaction steps via the formation of Compound II. These $1e^-$ redox reactions produce radical species (such as the tyrosyl radical when Tyr is the substrate)

conventional MPO activity assays, some assays are not feasible for these studies, and in others fine-tuning of the conditions is necessary. In isolated protein systems, we describe a modified TMB protocol to study the effects of sulfide on the peroxidase and chlorinating activities of the enzyme.

We also present two additional methods that may help to better understand the mediatory roles of sulfide in inflammatory processes in association with its interactions with MPO. The inhibitory potential of sulfide on the peroxidase activity of MPO from neutrophil homogenates can be assessed by the method of Bradley [30] with modifications. In this method a 50-fold dilution of the homogenates is necessary to obtain appropriate absorbance values during the spectrophotometric measurement in a 1 cm path length cuvette. However, due to the reversible nature of sulfide inhibition on the enzymatic activities of MPO, this dilution step can also dilute out the inhibitory effect of sulfide. This phenomenon has important implications on experiments, where MPO activity measurements are used to gain insight into various inflammatory processes and the roles of sulfide in these conditions.

Because the majority of MPO that is associated with detrimental ROS production in cardiovascular diseases is expressed by neutrophil granulocytes on site [21], measurements to assess the effects of sulfide on the in situ ROS-generating potential of neutrophils could be crucial to get deeper insights in these processes. Therefore, we developed a protocol to measure the effects of sulfide on the MPO peroxidase activity of live neutrophils. This method utilizes the ability of MPO to use tyrosine as a substrate in its peroxidase cycle to generate a fluorescent dityrosine product, which can be followed by spectrofluorometry [31, 32].

2 Materials

2.1 Common Stock Solutions and Reagents

1. N,N-Dimethylformamide (DMF), 100 v/v% liquid organic solvent.

2. Diethylenetriaminepentaacetic acid (DTPA) solid reagent, dissolve in the appropriate assay buffer.

3. 20 mM 3,5,3',5'-tetramethylbenzidine (TMB) stock solution, prepare by dissolving solid reagent in 100 v/v% DMF, protect from light, and store at 4 °C until use.

4. Phosphate-buffered saline (PBS), dissolve one tablet in 200 ml water containing 100 μM DTPA (this will give 10 mM phosphate buffer, 2.7 mM potassium chloride, and 137 mM sodium chloride at pH 7.4).

5. 30 w/w% hydrogen peroxide (Sigma).

6. 100 mM sodium azide stock solution, prepare by dissolving solid reagent in PBS, aliquot, and store at −80 °C until use.

7. 10 mM phosphate (KH_2PO_4) buffer (pH 7.4) containing 100 μM DTPA.

2.2 Preparation of Hydrogen Sulfide Stock Solution

1. Degassed ultrapure water.

2. 100 mM phosphate buffer (mix 81 v/v% 100 mM K_2HPO_4 with 19 v/v% 100 mM KH_2PO_4), pH 7.4.

3. 10 mM 5,5-dithio-bis-(2-nitrobenzoic acid) (DTNB) stock solution, make by dissolving solid reagent in 100 mM phosphate buffer (pH 7.4).

4. For the sulfide stock solution, dissolve $Na_2S \times 9H_2O$ crystals in ultrapure water as indicated in Subheading 3.1, steps 1–6.

2.3 Samples

1. 1 mg/ml myeloperoxidase enzyme (MPO) stock solution, prepare by dissolving powder in ultrapure water (see Note 1).

2. Isolated neutrophil granulocytes (homogenate or activated live cells), all experimentation involving human subjects must be conducted according to regional ethical regulations.

2.4 Measuring Sulfide Inhibition on the Peroxidase Activity of Purified MPO

1. Peroxidase assay buffer: 100 mM potassium phosphate buffer, pH 5.4 (mix 81 v/v% 100 mM K_2HPO_4 with 19 v/v% 100 mM KH_2PO_4), containing 100 μM DTPA and 8 v/v % DMF.

2.5 Measuring Sulfide Inhibition on the Chlorinating Activity of Isolated MPO

1. Chlorination assay buffer: 10 mM PBS (pH 7.4, 137 mM NaCl, 2.7 mM KCl, 10 mM phosphate buffer) made by dissolving one tablet of PBS in 200 ml ultrapure water containing 100 μM DTPA.

2. TMB developing reagent: 2 mM TMB in 400 mM sodium-acetate buffer (pH 5.4) containing 100 μM DTPA and 10 mM potassium iodide, prepare daily, and protect from light.

3. 1.5 M L-glycine stock solution, prepare daily by dissolving solid reagent in the chlorination assay buffer.

4. Sodium hypochlorite solution in 0.1 M NaOH (bleach).

2.6 The Effect of Sulfide on the MPO Peroxidase Activity in Neutrophil Homogenates

1. Hexadecyltrimethylammonium bromide (HTAB), solid cationic detergent.

2. 100 mM phosphate buffer (pH 6.0) containing 0.5 w/v % HTAB.

2.7 The Effect of Sulfide on the MPO Peroxidase Activity in Activated Live Neutrophil Samples

1. Hank's Balanced Salt Solution (HBSS), modified with sodium bicarbonate, without phenol red, liquid, sterile-filtered, suitable for cell culture (Lonza).

2. Dulbecco's phosphate-buffered saline (DPBS), 9.5 mM phosphate without calcium and magnesium, sterile-filtered, suitable for cell culture (Lonza).

3. Dextran from *Leuconostoc mesenteroides* (Sigma): 5 w/v% dissolved in PBS, prepare freshly, and sterilize by filtering on a 0.2 μm pore size sterile filter (Merck).

4. Histopaque reagent (Sigma).

5. NaCl: prepare hypotonic 0.2 w/v% or hypertonic 1.6 w/v% stock solutions by dissolving the adequate amounts of salt in ultrapure water, sterilize by filtering on a 0.2 μm pore size sterile filter, and store at 4 °C.

6. Cell counting dye, e.g., Trypan Blue Solution, 0.4%, Thermo Fisher Scientific.

7. 4 mM L-tyrosine stock solution, dissolve solid reagent in HBSS freshly, and sterilize by filtering on a 0.2 μm pore size sterile filter (*see* **Note 2**).

8. 100 mM L-methionine stock solution, dissolve solid reagent in HBSS freshly, and sterilize by filtering on a 0.2 μm pore size sterile filter.

9. Dimethyl sulfoxide (DMSO), sterilize by filtering on a 0.2 μm pore size sterile filter suitable for organic solvents.

10. 50 μg/ml phorbol 12-myristate 13-acetate (PMA) stock solution, prepare by dissolving solid reagent in sterile DMSO, aliquot, and store at −80 °C.

11. 1 mg/ml superoxide dismutase (SOD) stock solution, prepare by dissolving powder in HBSS, sterilize by filtering on a 0.2 μm pore size sterile filter, aliquot, and store at −80 °C (Sigma).

12. 1 mg/ml horseradish peroxidase (HRP) stock solution, prepare by dissolving powder in HBSS freshly, and do not freeze (Sigma).

13. 1 mg/ml stock bovine catalase stock solution, make by dissolving in HBSS freshly, and do not freeze (Sigma).

2.8 Instrumentation

1. UV-Vis micro-volume spectrophotometer, plate reader.

2. Spectrofluorometer, plate reader.

3. Tissue/cell homogenizer (Sartorius).

4. Syringe filter unit, 0.2 μm pore size, sterile (Merck).

5. Dry block heating thermostat.

6. Sterile hood.

7. Fume hood.

3 Methods

3.1 Preparation of Hydrogen Sulfide Stock Solution

Each stock solution should be prepared fresh daily and stored on ice before use. Use tightly closed and covered plastic bottles to avoid volatilization, oxidation, or interference of photo-induced reactions. For safety purposes, handle all working and stock solutions of sulfide under chemical fume hood and confirm appropriate air ventilation. Leftover sulfide solutions should be discarded into alkaline sodium hypochlorite solution.

1. Prepare DTNB stock solution: 4 mg/ml (~10 mM) in 100 mM phosphate buffer at pH 7.4.

2. Take a relatively large crystal of $Na_2S \times 9H_2O$, and rinse it repeatedly with ultrapure water. Discard the supernatant into sodium hypochlorite solution (*see* **Note 3**).

3. Dissolve the washed crystals in 5–6 ml degassed ultrapure water (*see* **Note 4**). Use this concentrated stock solution within 1–2 h.

4. Dilute the stock solution with ultrapure water, and measure the absorbance at 230 nm in quartz cuvette. For accurate determination of the concentration, sulfide needs to be in its single protonated HS^- form; therefore the pH of the solution has to be >9. To ensure this pH while avoiding the addition of extra base or buffer, the concentration of the diluted HS^- solution should be ~50–60 µM [17]. This concentration range should provide absorbance values between 0.4 and 0.5 at 230 nm in a 1 cm path length cuvette, because the molar absorption coefficient of HS^- at 230 nm is $7700 \ M^{-1} \ cm^{-1}$. Using this extinction coefficient, calculate the concentration of the diluted solution.

5. Add 1/10 volume unit of 10 mM DTNB to the sulfide solution in the same cuvette. An intense yellow product 2-nitro-5-thiobenzoate (TNB) is formed in 1–2 min. Measure absorbance at 412 nm (absorption maximum of TNB), and calculate the concentration of sulfide using $\varepsilon = 14{,}100 \ M^{-1} \ cm^{-1}$ and considering the fact that 2 mol equivalents of TNB are produced per sulfide under these conditions [33]. Use a dilution factor to correct for the addition of the DTNB aliquot.

6. Use the average value obtained from the two methods (under 4 and 5) as the applied concentration of the sulfide stock solution. The difference between the two calculated concentrations should be less than five percent, otherwise repeat the whole procedure with fresh reagents (*see* **Note 5**).

3.2 Steady-State Kinetic Measurements for the Peroxidase Activity of Isolated MPO in the Presence of Sulfide

In the peroxidase cycle, MPO catalyzes the oxidation of 3,5,3′,5′-tetramethylbenzidine (TMB) by hydrogen peroxide in two consecutive one-electron reactions. The formation of the blue product, the charge-transfer TMB diamine-diimine complex, can be followed by monitoring the absorbance at 650 nm [34, 35]. Because of the inherent reactivity of sulfide, we had to investigate its possible interference with the assay conditions by scavenging intermediate or product species (the TMB cation radical or diamine-diamine complex). We demonstrated that the concentration of the TMB diamine-diimine complex is stable in the presence of all applied concentrations of sulfide [19]. In the presence of 1 µM sulfide, the maximum measured values of the absorbance at 652 nm were congruous with the values that were measured in the absence of sulfide. Although the reaction rate was lower in the presence of sulfide, these results suggest that sulfide does not react with any of the intermediates but inhibits the peroxidase activity of MPO (*see* Fig. 1). In addition, the inhibitory effect of sulfide was followed by measuring the decrease in the concentration of hydrogen peroxide

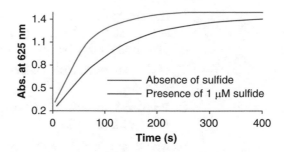

Fig. 1 Sulfide does not interfere with the assay products and intermediate species. The peroxidase activity of 6 nM MPO was measured by the TMB assay in the absence (red line) and the presence of 1 μM sulfide (blue line). Upon total depletion of the added 30 μM hydrogen peroxide, the absorbance changes were similar in the presence and absence of 1 μM sulfide, which indicates that sulfide does not react with TMB oxidation products. The figure was reproduced from ref. 19 with permission

using a peroxide selective electrode. Based on the kinetic data, we found that the IC_{50} values that were calculated by following product formation (TMB oxidation) and substrate consumption (loss of peroxide) were approximately equal [19]. These findings further corroborate that the measured effects of sulfide indeed represent MPO inhibition and not interference with the assay.

Therefore, this modified TMB assay is appropriate to study the interactions of sulfide with the peroxidase activity of MPO. Below you can find an example under particular conditions on how to conduct this assay.

1. TMB assays in our laboratory were carried out at 25 °C.

2. Prepare a larger volume of 100 mM potassium phosphate buffer at pH 5.4 (100 ml) containing 100 μM DTPA, 8 v/v% DMF (*see* **Notes 6** and **7**). This is the assay buffer for measuring the peroxidase activity of MPO (*see* **Notes 8** and **9**).

3. Prepare hydrogen sulfide stock and working solutions as described in Subheading 3.1, **steps 1–6** Make working solutions at 10–1000 μM concentrations by diluting the stock solution with the peroxidase assay buffer.

4. Prepare 10 ml H_2O_2 stock solution by dilution of the 30 w/w% solution with ultrapure water in a covered plastic tube (use, e.g., 10 μl of the 30% stock solution, and dilute it to 10 ml). Measure its absorbance before every experiment at 240 nm. Calculate the concentration of the stock solution using $\varepsilon = 43.6 \ \mathrm{M^{-1} \ cm^{-1}}$ (*see* **Note 10**).

5. Prepare 1 mM peroxide working solution by diluting the 10 mM stock solution with the assay buffer.

6. Prepare a 20 mM TMB stock solution in 10 ml 100 v/v% DMF, and determine the concentration of TMB by measuring

the absorbance at 285 nm where the extinction coefficient is 2.1×10^4 M^{-1} cm^{-1} [34]. TMB stock solution should be prepared freshly before every experiment.

7. Make 10 mM PBS buffer as described in Subheading 2.1, **step 4**.

8. Prepare 1 mg/ml MPO stock solution in ultrapure water (*see* **Notes 11** and **12**).

9. Make appropriate working solutions of MPO (200 nM) by diluting the stock solution with 10 mM phosphate (K_2HPO_4/KH_2PO_4) buffer (pH 7.4) containing 100 μM DTPA (*see* **Note 13**).

10. Prepare reaction mixture by mixing 50 μl TMB stock solution (final [TMB] = 1 mM), 6 μl MPO (final [MPO] = 6 nM), 50 μl sulfide solution (final [sulfide] = 0.5–50 μM), and 794 μl peroxidase assay buffer into a 1 ml quartz cuvette. For the control sample (no sulfide), use 50 μl of peroxidase assay buffer instead of the sulfide solution.

11. Start the reaction by adding 100 μl H_2O_2 working solution to the sample (final [H_2O_2] = 10 μM), and start the kinetic measurement immediately.

12. Measure the absorbance values at 652 nm every 2 s to follow the formation of the blue TMB oxidation product (*see* **Notes 14** and **15**).

13. Use GraphPad software (Prism, La Jolla, California) to analyze the measured data. After the measurement convert the absorbance values to product concentrations using the molar extinction coefficient of the TMB diamine-diimine complex ($\varepsilon_{653 \text{ nm}} = 3.9 \times 10^4$ M^{-1} cm^{-1}). Use the linear sections of the time/concentration traces to calculate the initial reaction rates from the slopes of the linear regression analyses. Use the initial reaction rate of the untreated sample (without sulfide) as 100%, and calculate the percentage of inhibition for all sulfide concentrations (*see* Fig. 2a). Construct a dose-response curve by plotting the percentage inhibition as a function of log[sulfide]. Fit the response curve with a sigmoidal function. The IC_{50} value (representing 50% inhibition of the enzymatic activity by sulfide) can be obtained from the inflection point of the fitted curve (*see* Fig. 2b).

3.3 Semiquantitative Measurement for the Effect of Sulfide on the Chlorinating Activity of MPO

Due to interference of sulfide with conventional assays that use ascorbic acid [36] or NADPH [37] to trap MPO-produced HOCl, these methods cannot be used to study the effects of sulfide on the chlorinating activity of MPO (*see* **Note 16**). The published TMB method to measure the chlorinating activity of MPO is based on the following reaction sequence: HOCl, which is produced by MPO (*see* Scheme 1), reacts with taurine to generate a taurine-

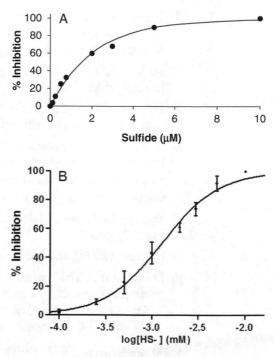

Fig. 2 The inhibitory effect of sulfide on the peroxidase activity of MPO. (**a**) Representative curve for the inhibition of the peroxidase activity of 6 nM MPO by sulfide at 30 μM H_2O_2. Curve shows the percentage of inhibition as a function of added sulfide (0.1–10 μM) concentration. (**b**) A dose-response plot (data points) can be constructed by applying a logarithmic transformation to sulfide concentrations. The IC_{50} concentration of sulfide (at this concentration the enzymatic activity is inhibited by 50%) can be obtained from the inflection point of the fitted sigmoidal function to the data (black line). Figure **a** was reproduced from ref. 19 with permission

monochloramine species, which then oxidize iodide to hypoiodous acid. Hypoiodous acid reacts with TMB to produce the colored diamine-diamine species that can be measured spectrophotometrically at 652 nm [38]. However, because the reaction of sulfide with HOCl has a rate constant representing nearly diffusion-controlled rates (2×10^9 M/s) [18], we had to modify the assay conditions to avoid HOCl scavenging by sulfide. To efficiently trap the majority of the produced HOCl as a monochloramine derivative, glycine at 1 M concentration had to be used (*see* **Note 17**). This modification had no effect on the measured halogenation activity of MPO in the absence of sulfide. The following examples give guidance on how to use this method:

1. Prepare 10 mM PBS buffer as described in Subheading 2.1, **step 4**.

2. Make a larger volume of chlorination assay buffer: 100 mM PBS and 100 μM of DTPA in 100 ml water (pH 7.4).

3. Prepare HS^- stock solution as described in Subheading 3.1, **steps 1–6**, and dilute with the chlorination assay buffer to obtain the corresponding working solutions (3–150 μM).

4. Make a 10 mM H_2O_2 stock solution as indicated in Subheading 3.2, **steps 4** and **5**.

5. Prepare a 200 μM H_2O_2 working solution from the stock solution by dilution using the chlorination assay buffer.

6. Dissolve 1.5 M glycine in 10 mM PBS containing 100 μM DTPA. For complete dissolution elevate the pH of the solution to 10–11, and vortex it thoroughly. After all the glycine was dissolved, set the pH to 7.4.

7. Prepare a 20 mM TMB stock solution as indicated in Subheading 3.2, **step 6**.

8. Prepare 100 ml 400 mM Na-acetate buffer (pH 5.4).

9. Dilute 200 μl TMB stock solution tenfold with the Na-acetate buffer, and dissolve 10 mM potassium iodide in it. This is the TMB developing solution containing 2 mM TMB, 10 v/v% DMF, and 100 μM potassium iodide (*see* **Note 18**).

10. Prepare MPO stock solution as described in Subheading 3.2, **steps 8** and **9** and working solution (200 nM) with 10 mM PBS containing 100 μM DTPA (*see* **Notes 12** and **13**).

11. Prepare the reaction mixture by adding 22.5 μl of 200 nM MPO (at a final concentration of 5 nM) with 147.5 μl of 100 mM phosphate buffer containing 137 mM sodium chloride and 600 μl glycine stock solution (at a final glycine concentration of 1 M).

12. Add 30 μl of sulfide from the appropriate sulfide working solution (*see* **step 3**) to obtain final sulfide concentrations in the range of 0.1–5 μM. Add 30 μl of the assay buffer for the control sample (no sulfide).

13. Start the reaction by adding 100 μl H_2O_2 from the working solution (*see* **step 5**) at a final concentration of 25 μM.

14. Take 80 μl aliquots from the sample at different time points (e.g., 30–150 s), and mix it with 8 μl of 100 mM sodium azide solution (final conc. = 1 mM) to stop the reaction (*see* **Note 19**).

15. Add 20 μl of TMB developing reagent to the samples, and incubate for 5 min during which TMB is oxidized by the generated glycine chloramine species via the catalytic actions of iodide (*see* **Note 20**).

16. To quantify the amount of generated HOCl, use standard HOCl solutions with known concentrations (in the range of 5–100 μM), and carry out **steps 14** and **15** with these standard samples. Develop a calibration curve by plotting the generated

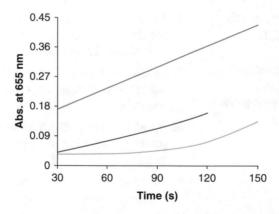

Fig. 3 Effect of sulfide on the chlorinating activity of MPO. The initial phase of the kinetic curves indicates sulfide-mediated inhibition of the chlorinating activity of MPO. However, at later time points the enzymatic activity is slowly recovered due to sulfide consumption by MPO-generated glycine chloramine species. Kinetic curves are representative of traces that were recorded in the absence of sulfide (blue line), in the presence of 1 μM (red line), or 4 μM (green line) sulfide. The figure was reproduced from ref. 19 with permission

TMB diamine-diamine concentrations as a function of the applied HOCl concentration (*see* **Note 21**).

17. Measure the absorbance values at 650 nm (we used 384-well microplates with a sample volume of 100 μl) (*see* **Notes 14** and **15**).

18. Data analyses should be carried out as for the peroxidative activity assay (*see* Subheading 3.2, **step 13**) but only using early time points of the kinetic traces (before significant sulfide is consumed by glycine chloramine, *see* below).

The following limitation of this modified assay must be taken into account to avoid misinterpretation of the data: We now know that sulfide also reacts with chloramines very efficiently; therefore only early time points can be used from this assay, and the results should be handled as semiquantitative regarding the concentrations of the TMB diamine-diamine species produced by MPO. Although the early time points reflect the inhibitory potential of sulfide on the chlorinating activity of MPO, the enzymatic activity gradually returned to its original (no sulfide) state over time (*see* Fig. 3). This is the result of the loss of sulfide via oxidation by the generated glycine chloramine. As we demonstrated before, sulfide-mediated inhibition of MPO is fully reversible, and the generated sulfide oxidation products express no inhibitory potential on MPO activity [19].

3.4 Measurements for the Effects of Sulfide on the Peroxidase Activity of MPO in Human Neutrophil Preparations

1. Isolate neutrophil granulocytes from fresh blood, collect in sterile EDTA collection tubes, and store them on ice until use (*see* **Notes 22** and **23**).

2. Bring the temperature of the samples to 37 °C before treatment to create a more physiologically relevant environment for the cells.

3. Sterilize all reagents and solutions by filtration, and use sterile plastic equipment to prevent uncontrolled activation of neutrophils (*see* **Note 24**).

4. Conduct the experiments under a sterile hood to avoid contamination.

3.4.1 Isolation of Neutrophil Granulocytes

1. Prepare 5 w/v% dextran solution with PBS (*see* **Note 25**).

2. Prepare a hypotonic 0.2 w/v% NaCl solution for the lysis step and a hypertonic 1.6 w/v% NaCl solution to restore the osmotic strength of the buffer after the hypotonic lysis.

3. Pipette 2.5 ml of dextran (1 w/v% final concentration) to 10 ml of blood and mix gently. Vigorous shaking can cause red blood cells to disintegrate and elevate the dissolved oxygen content of the sample. These events could interfere with detections of ROS-generating activities of neutrophil cells (*see* **Note 26**).

4. Incubate the mixture for 30–40 min at room temperature.

5. Red blood cells should settle, and the pale yellow supernatant can be separated. This supernatant contains the leukocyte fraction of the blood.

6. Discard the pellet of red blood cells.

7. Centrifuge the supernatant at $500 \times g$ for 5 min.

8. Discard the supernatant and resuspend the cells in 5 ml PBS.

9. After the removal of the remaining dextran, use a Histopaque gradient to separate the polymorphonuclear cells from the mononuclear fraction. Pipette the cell suspension on 5 ml Histopaque (1.077 g/ml) reagent slowly and carefully without mixing the solutions.

10. Centrifuge the cells at $1000 \times g$ for 30 min without using the break function of the centrifuge. Under these conditions, mononuclear cells should stay on the layer between the two liquid phases and polymorphonuclear cells settle on the bottom of the tube (*see* **Note 27**).

11. Discard the supernatant, and resuspend the cells in 10 ml PBS.

12. Centrifuge at $500 \times g$ for 5 min.

13. Discard the supernatant, and resuspend the cells in 5 ml 0.2 w/v% NaCl solution for 1 min to lyse the remaining red blood cells (*see* **Note 28**).

14. Add 5 ml 1.6 w/v% NaCl solution to the sample to restore isotonic conditions (*see* **Note 29**).

15. Wash the cells with PBS as described in **steps 7** and **8**, and resuspend the cells in 1 ml HBSS.

16. Prepare appropriate dilutions for cell counting (e.g., 10^5 cells/ml).

17. Use Trypan Blue count viable neutrophils and other contaminating cells. The purity of the isolated neutrophils should be ~98% to ensure reproducibility of experiments carried out with these cells. The purity of the preparation can be verified by Giemsa staining and morphological analysis [39] (*see* **Note 30**).

18. Dilute the sample with HBSS to ~10^7 cells/ml concentration, and store the cell suspension on ice until use.

3.4.2 Effects of Sulfide on the Peroxidase Activity of MPO in Human Neutrophil Homogenates

The method previously published by Bradley et al. [30] combined with a modified TMB assay can be used to assess MPO activity in the presence of sulfide in human neutrophil homogenates. An important modification is that in order to avoid the loss of sulfide's inhibitory effect, neutrophil homogenates should be diluted to the appropriate assay conditions before the addition of sulfide when the inhibitory potential of sulfide is tested [19] (*see* Fig. 4 and **Note 31**). On the other hand, if the measurement of MPO activity in a particular biological sample is aimed at assessing the intensity of inflammatory processes (accumulation of neutrophils is a hallmark of inflammatory diseases [40, 41] in sulfide-treated samples), then it is important to use an appropriate dilution factor during the assay to dilute out the inhibitory effect of sulfide on MPO activity (which in this case would confine the results).

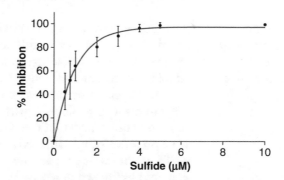

Fig. 4 Sulfide's effect on MPO peroxidase activity in neutrophil homogenates. Percentage of inhibition of MPO peroxidase activity by various concentrations of added sulfide in neutrophil lysates. Sulfide was added after the dilution of supernatant of the homogenates to the assay conditions. The figure was reproduced from ref. 19 with permission

1. Prepare hydrogen sulfide and hydrogen peroxide stock solutions as described earlier in Subheadings 3.1, **steps 1–6** and 3.2, **steps 4** and **5**, and make appropriate dilutions in the peroxidase assay buffer.

2. Isolate the neutrophils using the protocol that was described in Chap. 3.4.1, **steps 1–18**.

3. Use 100 mM phosphate buffer containing 0.5 w/v% HTAB (pH 6), to resuspend the neutrophils at the final step of isolation. Further dilute the cell suspension with the phosphate buffer if needed (*see* **Note 8**). Relatively high sample volume (~2–3 ml) is needed to carry out proper mechanical lysis. Cell concentration should be ~5 × 10^5 cells/ml.

4. Use tissue/cell culture homogenizer for 1–2 min to disrupt the isolated neutrophils (*see* **Note 32**).

5. Centrifuge the sample at 40,000 × g for 15 min.

6. Discard the remaining cell debris.

7. Pipette the supernatant into a new tube.

8. Mix 20 μl supernatant, 870 μl peroxidase assay buffer, and 50 μl TMB solution (to achieve a final TMB concentration of 1 mM).

9. Pipette the sample into a 1 ml quartz cuvette (with 1 cm path length), and add 40 μl of the appropriate sulfide working solution (to achieve a final sulfide concentration in the range of 0.5–10 μM) or 40 μl assay buffer for the control sample (no sulfide), and incubate for 5–7 min.

10. Start the reaction by the addition 20 μl of H_2O_2 (final $[H_2O_2] = 25$ μM).

11. Measure the absorbance at 652 nm (*see* **Note 14**) every 2 s for ~15 min.

12. Analyze the data as indicated in Subheading 3.2, **step 13**.

3.4.3 Measurement of MPO Peroxidase Activity of Activated Live Neutrophils in the Presence of Sulfide

Conventional assays are not suitable because they work at nonphysiological conditions or sulfide interferes with them. For example, the TMB assay is carried out under acidic conditions that are not suitable for experiments with live neutrophils. Therefore, a new method was elaborated to ensure physiological conditions for the cells and avoid interference with sulfide. The method is based on the following sequence of events: Upon stimulation of neutrophils in the presence of tyrosine, MPO—that is excreted from the azurophilic granules—catalyze the oxidation of tyrosine by NOX2-generated hydrogen peroxide to give tyrosyl radicals. Tyrosyl radicals then combine to form dityrosine, which is a stable product that can be measured by spectrofluorometry (*see* Scheme 3).

1. Prepare 15–20 ml of 4 mM L-tyrosine solution as indicated in Subheading 2.7, **step 7** (*see* **Notes 33** and **34**).

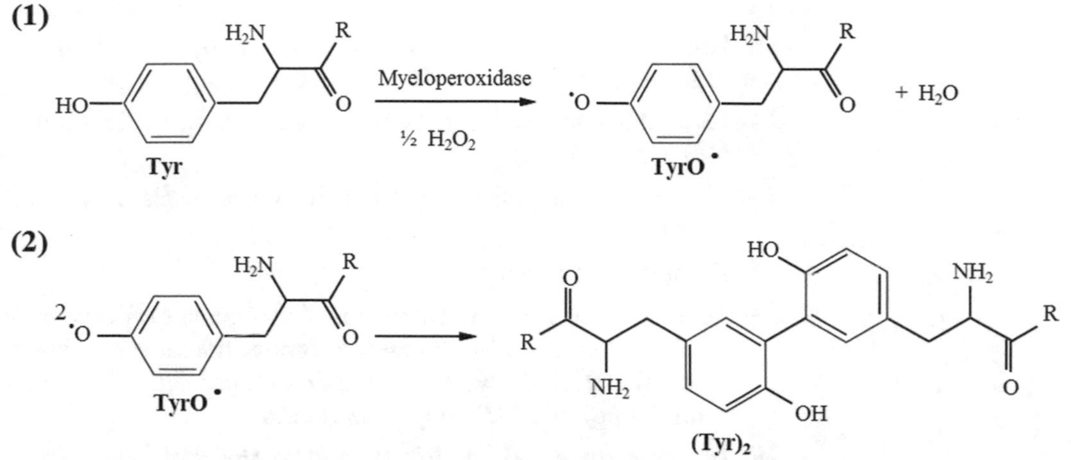

(1)

(2)

Scheme 3 Myeloperoxidase-catalyzed formation of dityrosine. (Reaction 1) Myeloperoxidase catalyzes the one-electron oxidation of tyrosine by H_2O_2 to give tyrosyl radical species (this phenoxyl radical species has different resonance structures; in the most abundant form, the unpaired electron is delocalized on the phenol ring) via the peroxidase cycle (*see* Scheme 2). (Reaction 2) Two tyrosyl radicals combine to form covalently bound fluorescent dimers in a very fast reaction ($k \sim 5 \times 10^8 \ M^{-1} \ s^{-1}$)

2. Prepare 15–20 ml of 100 mM L-methionine solution in HBSS. Methionine can efficiently capture hypochlorous acid, thus protects the neutrophils and MPO from HOCl-induced damage [42, 43].

3. Prepare a 50 μg/ml PMA stock solution in DMSO (*see* **Note 35**).

4. Make 1 mg/ml catalase and 1 mg/ml SOD stock solutions in HBSS (*see* **Note 36**).

5. Prepare the reaction mixtures in sterile tubes: 100 μl neutrophil sample (final concentration = 2×10^6 cells/ml), 80 μl HBSS buffer, 250 μl L-tyrosine (final [Tyr] = 2 mM), 50 μl L-methionine (final [Met] = 10 mM), and 10 μl SOD (final [SOD] = 20 μg/ml).

6. Equilibrate the temperature of all reaction mixtures at 37 °C in a heat block system.

7. Add 10 μl of working sulfide solution with the appropriate concentration to the samples (final [sulfide] = 5–50 μM), and incubate for 5 min.

8. Activate the neutrophil granulocytes by adding 1 μl PMA stock solution (final [PMA] = 100 ng/ml) (*see* **Note 37**) or add 1 μl DMSO to the control sample (no sulfide), and incubate the samples at 37 °C for 20 min (*see* **Note 38**).

9. During the incubation gently rotate the samples in every 3–5 min (*see* **Note 39**).

10. After 20 min add 10 μl of catalase stock solution (final [catalase] = 20 μg/ml) to stop the reaction by quenching the residual hydrogen peroxide, and place the samples on ice.

11. Cool the samples for 5 min followed by centrifugation at 500 × g for 5 min.

12. Pipette 100 μl of supernatant into the wells if a black microplate.

13. Discard the cell pellet.

14. To determine the concentration of the generated dityrosine product, prepare cell-free standards under the same conditions by oxidizing 2 mM tyrosine by hydrogen peroxide (5–50 μM) using 20 μg/ml HRP enzyme as catalyst.

15. Incubate the standards for 10 min in the dark, and pipette them into the wells of the black plate along with the samples.

16. Measure the fluorescence at 325 nm emission and 400 nm extinction wavelengths using a fluorescent plate reader. Calculate the dityrosine concentration of the samples considering that 1 mM H_2O_2 is used to produce 1 mM dityrosine (*see* **Note 40**).

17. Use GraphPad software (Prism, La Jolla, California) to analyze the measured data. Use the relative fluorescence unit (RFU) values measured after 20 min of the untreated sample as 100%, and calculate the percent inhibition for every sulfide concentration. Follow the instruction described in Subheading 3.2, **step 13**, and construct a dose-response curve to determine the IC_{50} value.

4   Notes

1. We used MPO and neutrophils derived from human donors, but samples from other species are suitable as well for these experiments.

2. In order to dissolve all the tyrosine, the pH of the solution has to be raised to 12. After complete dissolution of the solid chemical, lower the pH back to 7.4. Measure the absorbance at 280 nm, and calculate the concentration using the following molar extinction coefficient $\varepsilon_{Tyr} = 1280$ M^{-1} cm^{-1} (*see* **Notes 33** and **34**).

3. Concentrations of sulfide stock solutions can't be assumed based on weighing out the corresponding solid sulfide salt because of apparent contaminations in the commercially available chemicals. Please consult a previous report from our laboratory for a more detailed description of sulfide salts and a proposal on how to make and handle sulfide solutions [17].

4. Degassing refers to bubbling with an inert gas (argon or nitrogen) or to sonication at low pressure in order to reduce the concentration of dissolved oxygen in the solution and prevent unwanted sulfide oxidation.

5. The most likely reason for an unacceptable difference ($>5\%$) in the measured sulfide concentration by the two methods is the formation of sulfide oxidation products during the aging of the Na_2S salt. In most cases using a new batch of sulfide salt resolves the problem.

6. Potassium phosphate is better to use in experiments with isolated MPO instead of sodium phosphate because it is less contaminated with metal ions; however it is still advised to use 100 μM DTPA to chelate trace metal ion contamination.

7. TMB has a lower solubility in water-based buffer; therefore we used 8 v/v% DMF in all the TMB assay buffers to improve the solubility of TMB.

8. MPO is a sticky protein that easily adheres to plastic surfaces of the pipette tips or tubes. Specifically in low, nanomolar concentrations, this could lead to significant loss of activity during the assay. Use HTAB, a cationic detergent to avoid loss of MPO. At higher concentrations of MPO, it is better to avoid the addition of detergent.

9. Make sure that the buffer (or any other solutions) does not contain chloride, because the presence of chloride would result in the generation of HOCl via the chlorination cycle (*see* Scheme 1 and Subheadings 1 and 3.3) and confine the measurements.

10. The 10 mM H_2O_2 solution can be stored at 4 °C in a covered plastic container; however the concentration should be checked before every experiment.

11. Highly purified MPO with a purity index (A_{428}/A_{280}) of at least 0.85 was purchased as lyophilized powder from Planta Natural Products (http://www.planta.at). After preparing a fresh stock solution, check the concentration by measuring the absorbance at 430 nm ($\varepsilon = 91,000$ M^{-1} cm^{-1} per heme).

12. Check the peroxidase activity of the dissolved enzyme with the TMB peroxidase assay without added sulfide before every experiment. Repeat this step every day after all the experiments as well to be able to verify that the measured decrease of activity comes from the inhibitory effect of sulfide.

13. Dilute MPO working solutions (e.g., 200 nM) with 10 mM K_2HPO_4/KH_2PO_4 buffer that does not contain any chloride. It prevents the production of HOCl by MPO that could damage the enzyme.

14. Measure the time that is passed between the start of the reaction and the start of the measurement, and correct your timescale according to it.

15. We used a BioTek PowerWave plate reader for the kinetic measurements, but other spectrophotometric devices are also suitable.

16. Considering the extremely rapid reaction of sulfide with HOCl [18], we investigated whether we can adopt these methods to measure the effects of sulfide on the halogenation cycle of MPO. Based on kinetic calculations, the required concentrations of these trapping agents (NADH and ascorbate) to outcompete the reaction between sulfide and HOCl would not allow reliable spectrophotometric analyses of the enzymatic reactions.

17. Calculations based on published rate constants suggested that the presence of 1 M amine is required to outcompete the reaction of sulfide with HOCl under the applied conditions [18, 44]. Glycine was chosen instead of taurine because it has higher solubility in aqueous buffers and it is less contaminated with metal ions.

18. Prepare TMB and glycine stock solution as well as the developing reagent freshly every day. Use the developing reagent within an hour to prevent the autoxidation of potassium iodide, or make a fresh solution if necessary.

19. Using catalase at a final concentration of 20 μg/ml could also be applied to quench H_2O_2 in the reaction mixture and stop the reaction [38]. However, sulfide also interacts with the activity of catalase [45], and therefore the use of azide is recommended.

20. Relatively high concentration of potassium iodide is required for quantitative oxidation of TMB [38].

21. The samples and standards should be kept on ice and assayed within 30 min because after that time a significant decline in absorbance was reported. Keep all standard solutions on ice until measurement [38].

22. Cooling isolated neutrophils would mostly prevent autoactivation processes and thereby dampen background production of reactive oxygen species. However, avoid freezing the cells because that could cause the loss of their ROS-producing functions.

23. Experiments with neutrophil preparations should be carried out within 2–4 h after separation [46, 47].

24. It is crucial to use endotoxin-free reagents and sterile-filtered solutions to prevent the activation of phagocytic pathways in the cells. Neutrophils also reported to be sensitive to glass and

polystyrene equipment, but polytetrafluoroethylene or polypropylene does not interfere with their functions [48, 49].

25. Since red blood cells have limited response to environmental exposure, they are more sensitive to osmotic changes. In dextran solution they would settle on the bottom of the tube more rapidly than white blood cells; hence they can be easily separated.

26. Rough shaking causes mechanical damage to neutrophils and leads to their activation and/or aggregation. This can cause a lower yield of viable cells at the end of the isolation.

27. The separation of mononuclear cells from polymorphonuclear cells can alternatively be performed on a Percoll gradient [50]. We did not find any differences in the effectiveness of the two methods or the viability of the isolated cells.

28. Although neutrophils are more resistant to hypotonic conditions than red blood cells, it is important to keep the lysis step short (not longer than 1 min), to protect the neutrophils from eventual damage and lysis.

29. If further removal of red blood cells is needed, repeat this step (not more than once). However, do not incubate with the lysis buffer longer than 1 min to avoid the lysis of granulocytes. After appropriate lysis, the pellet on the bottom of the tube should be white, without any remaining red blood cells.

30. With Trypan Blue Solution, viable cells can be counted; however it is not suitable to identify neutrophils. Use other cell-penetrating DNA dyes (e.g., Giemsa) to verify the purity of the preparation. Notably, this protocol for isolating neutrophils is a well-adjusted method to separate the granulocytes from red blood cells and other leukocytes.

31. This is important because sulfide inhibition is reversible and excessive dilution results in its dissociation from the heme center of MPO and the loss of its inhibitory power on the peroxidase activity [19].

32. We used Potter S Homogenizer for these experiments, but other mechanical tissues or cell culture homogenizers can also be used at this step. This method provides the appropriate disruption of the cells and its granules.

33. Tyrosine has a relatively low solubility in water-based buffers at pH 7.4. Make sure that all of the solid material is dissolved before measuring the concentration of the solution. An indication of a non-complete dissolution is an upward shift in the baseline of the UV-Vis spectrum due to light scattering induced by colloidal particles.

34. L-Tyrosine stock solution can be stored in −80 °C. Freeze and thaw cycles can influence the solubility of the reagent; therefore

check the concentration every time, and prepare a fresh solution if necessary.

35. PMA is a carcinogen; therefore use extreme caution when handling this compound, and follow the personal protection protocol described in the data sheet provided by the vendor [51, 52].

36. SOD catalyze the dismutation of superoxide to hydrogen peroxide which is a key substrate for MPO in this measurement. The addition of SOD is necessary to avoid interfering reactions of tyrosyl radicals [53, 54] and MPO [55, 56] with superoxide.

37. There are other suitable stimulants that can be used instead of PMA, for example, fMet-Leu-Phe (fMLP) [57].

38. PMA is soluble and stable in organic solvents; therefore we used DMSO which is compatible at these concentrations with in vitro studies of neutrophil cells. However, it is important that the final concentration of DMSO should not exceed 0.5 v/v% to avoid its toxicity [58].

39. Gentle rotation of the tubes would keep the cells in suspension and aerated. This way the aggregation of the cells or high local concentrations of the reaction products and intermediates can be avoided.

40. Laser flash photolysis experiments indicated that the reaction rate of tyrosyl radicals and sulfide is negligible during the 20-min incubation period; therefore we concluded that sulfide does not interfere with the assay by capturing the tyrosyl radical intermediate species.

Acknowledgments

This work was supported by The Hungarian National Science Foundation (OTKA; grant No.: K 109843, KH17_126766 and K18_129286 for P.N. and K 112333 for J.B.) and the National Institutes of Health (grant No.: R21AG055022-01 for P.N.). Financial supports from the Hungarian Government in a GINOP-2.3.2-15-2016-00043 project (for J.B. and P.N.) and from the European Union in a European Regional Development Fund are also acknowledged. The research group is supported by the Hungarian Academy of Sciences (11003). P.N. is a János Bolyai Research Scholar of the Hungarian Academy of Sciences. Dojindo Molecular Technologies Inc. is greatly acknowledged for their kind support of high-quality chemical supplies.

References

1. Szabo C (2007) Hydrogen sulphide and its therapeutic potential. Nat Rev Drug Discov 6 (11):917–935. https://doi.org/10.1038/nrd2425

2. Wang R (2012) Physiological implications of hydrogen sulfide: a whiff exploration that blossomed. Physiol Rev 92(2):791–896. https://doi.org/10.1152/physrev.00017.2011

3. Kimura H (2014) Production and physiological effects of hydrogen sulfide. Antioxid Redox Signal 20(5):783–793. https://doi.org/10.1089/ars.2013.5309

4. Wagner F, Asfar P, Calzia E et al (2009) Bench-to-bedside review: hydrogen sulfide—the third gaseous transmitter: applications for critical care. Crit Care 13(3):213. https://doi.org/10.1186/cc7700

5. Jha S, Calvert JW, Duranski MR et al (2008) Hydrogen sulfide attenuates hepatic ischemia-reperfusion injury: role of antioxidant and anti-apoptotic signaling. Am J Physiol Heart Circ Physiol 295(2):H801–H806. https://doi.org/10.1152/ajpheart.00377.2008

6. Kimura Y, Goto Y, Kimura H (2010) Hydrogen sulfide increases glutathione production and suppresses oxidative stress in mitochondria. Antioxid Redox Signal 12(1):1–13. https://doi.org/10.1089/ars.2008.2282

7. Kimura Y, Kimura H (2004) Hydrogen sulfide protects neurons from oxidative stress. FASEB J 18(10):1165–1167. https://doi.org/10.1096/fj.04-1815fje

8. Fu Z, Liu X, Geng B et al (2008) Hydrogen sulfide protects rat lung from ischemia-reperfusion injury. Life Sci 82 (23–24):1196–1202. https://doi.org/10.1016/j.lfs.2008.04.005

9. Whiteman M, Cheung NS, Zhu YZ et al (2005) Hydrogen sulphide: a novel inhibitor of hypochlorous acid-mediated oxidative damage in the brain? Biochem Bioph Res Commun 326(4):794–798. https://doi.org/10.1016/j.bbrc.2004.11.110

10. Laggner H, Muellner MK, Schreier S et al (2007) Hydrogen sulphide: a novel physiological inhibitor of LDL atherogenic modification by HOCl. Free Radic Res 41(7):741–747. https://doi.org/10.1080/10715760701263265

11. Nagy P (2015) Mechanistic chemical perspective of hydrogen sulfide signaling. Methods Enzymol 554:3–29. https://doi.org/10.1016/bs.mie.2014.11.036

12. Whiteman M, Armstrong JS, Chu SH et al (2004) The novel neuromodulator hydrogen sulfide: an endogenous peroxynitrite 'scavenger'? J Neurochem 90(3):765–768. https://doi.org/10.1111/j.1471-4159.2004.02617.x

13. Yonezawa D, Sekiguchi F, Miyamoto M et al (2007) A protective role of hydrogen sulfide against oxidative stress in rat gastric mucosal epithelium. Toxicology 241(1–2):11–18. https://doi.org/10.1016/j.tox.2007.07.020

14. Mani S, Li H, Untereiner A et al (2013) Decreased endogenous production of hydrogen sulfide accelerates atherosclerosis. Circulation 127(25):2523–2534. https://doi.org/10.1161/CIRCULATIONAHA.113.002208

15. Whiteman M, Haigh R, Tarr JM et al (2010) Detection of hydrogen sulfide in plasma and knee-joint synovial fluid from rheumatoid arthritis patients: relation to clinical and laboratory measures of inflammation. Ann N Y Acad Sci 1203:146–150. https://doi.org/10.1111/j.1749-6632.2010.05556.x

16. Elrod JW, Calvert JW, Morrison J et al (2007) Hydrogen sulfide attenuates myocardial ischemia-reperfusion injury by preservation of mitochondrial function. Proc Natl Acad Sci USA 104(39):15560–15565. https://doi.org/10.1073/pnas.0705891104

17. Nagy P, Palinkas Z, Nagy A et al (2014) Chemical aspects of hydrogen sulfide measurements in physiological samples. Biochim Biophys Acta 1840(2):876–891. https://doi.org/10.1016/j.bbagen.2013.05.037

18. Nagy P, Winterbourn CC (2010) Rapid reaction of hydrogen sulfide with the neutrophil oxidant hypochlorous acid to generate polysulfides. Chem Res Toxicol 23(10):1541–1543. https://doi.org/10.1021/tx100266a

19. Palinkas Z, Furtmuller PG, Nagy A et al (2015) Interactions of hydrogen sulfide with myeloperoxidase. Br J Pharmacol 172:1516–1532. https://doi.org/10.1111/bph.12769

20. Winterbourn CC, Kettle AJ, Hampton MB (2016) Reactive oxygen species and neutrophil function. Annu Rev Biochem 85:765–792. https://doi.org/10.1146/annurev-biochem-060815-014442

21. Nussbaum C, Klinke A, Adam M et al (2013) Myeloperoxidase: a leukocyte-derived protagonist of inflammation and cardiovascular disease. Antioxid Redox Signal 18 (6):692–713. https://doi.org/10.1089/ars.2012.4783

22. Klebanoff SJ (2005) Myeloperoxidase: friend and foe. J Leukoc Biol 77(5):598–625. https://doi.org/10.1189/jlb.1204697

23. Nicholls SJ, Hazen SL (2005) Myeloperoxidase and cardiovascular disease. Arterioscler Thromb Vasc Biol 25(6):1102–1111. https://doi.org/10.1161/01.ATV.0000163262.83456.6d

24. Stamp LK, Khalilova I, Tarr JM et al (2012) Myeloperoxidase and oxidative stress in rheumatoid arthritis. Rheumatology (Oxford) 51 (10):1796–1803. https://doi.org/10.1093/rheumatology/kes193

25. Heinecke JW (1997) Mechanisms of oxidative damage of low density lipoprotein in human atherosclerosis. Curr Opin Lipidol 8 (5):268–274

26. Klebanoff SJ (1968) Myeloperoxidase-halide-hydrogen peroxide antibacterial system. J Bacteriol 95(6):2131–2138

27. Furtmuller PG, Burner U, Obinger C (1998) Reaction of myeloperoxidase compound I with chloride, bromide, iodide, and thiocyanate. Biochemistry 37(51):17923–17930

28. vanDalen CJ, Whitehouse MW, Winterbourn CC et al (1997) Thiocyanate and chloride as competing substrates for myeloperoxidase. Biochem J 327:487–492

29. Furtmuller PG, Zederbauer M, Jantschko W et al (2006) Active site structure and catalytic mechanisms of human peroxidases. Arch Biochem Biophys 445(2):199–213. https://doi.org/10.1016/j.abb.2005.09.017

30. Bradley PP, Priebat DA, Christensen RD et al (1982) Measurement of cutaneous inflammation: estimation of neutrophil content with an enzyme marker. J Invest Dermatol 78 (3):206–209

31. Heinecke JW, Li W, Daehnke HL 3rd et al (1993) Dityrosine, a specific marker of oxidation, is synthesized by the myeloperoxidase-hydrogen peroxide system of human neutrophils and macrophages. J Biol Chem 268 (6):4069–4077

32. Heinecke JW, Li W, Francis GA et al (1993) Tyrosyl radical generated by myeloperoxidase catalyzes the oxidative cross-linking of proteins. J Clin Invest 91(6):2866–2872. https://doi.org/10.1172/JCI116531

33. Vasas A, Doka E, Fabian I et al (2015) Kinetic and thermodynamic studies on the disulfide-bond reducing potential of hydrogen sulfide. Nitric Oxide 46:93–101. https://doi.org/10.1016/j.niox.2014.12.003

34. Josephy PD, Eling T, Mason RP (1982) The horseradish peroxidase-catalyzed oxidation of 3,5,3',5'-tetramethylbenzidine. Free radical and charge-transfer complex intermediates. J Biol Chem 257(7):3669–3675

35. Marquez LA, Dunford HB (1997) Mechanism of the oxidation of 3,5,3',5'-tetramethylbenzidine by myeloperoxidase determined by transient- and steady-state kinetics. Biochemistry 36(31):9349–9355. https://doi.org/10.1021/bi970595j

36. Kettle AJ, Winterbourn CC (1994) Assays for the chlorination activity of myeloperoxidase. Methods Enzymol 233:502–512

37. Auchere F, Capeillere-Blandin C (1999) NADPH as a co-substrate for studies of the chlorinating activity of myeloperoxidase. Biochem J 343:603–613. https://doi.org/10.1042/0264-6021:3430603

38. Dypbukt JM, Bishop C, Brooks WM et al (2005) A sensitive and selective assay for chloramine production by myeloperoxidase. Free Radic Biol Med 39(11):1468–1477. https://doi.org/10.1016/j.freeradbiomed.2005.07.008

39. Barcia JJ (2007) The Giemsa stain: its history and applications. Int J Surg Pathol 15 (3):292–296. https://doi.org/10.1177/1066896907302239

40. Amulic B, Cazalet C, Hayes GL et al (2012) Neutrophil function: from mechanisms to disease. Annu Rev Immunol 30:459–489. https://doi.org/10.1146/annurev-immunol-020711-074942

41. Nauseef WM, Borregaard N (2014) Neutrophils at work. Nat Immunol 15(7):602–611. https://doi.org/10.1038/ni.2921

42. Peskin AV, Winterbourn CC (2001) Kinetics of the reactions of hypochlorous acid and amino acid chloramines with thiols, methionine, and ascorbate. Free Radic Biol Med 30 (5):572–579

43. Levine RL, Mosoni L, Berlett BS et al (1996) Methionine residues as endogenous antioxidants in proteins. Proc Natl Acad Sci USA 93 (26):15036–15040. https://doi.org/10.1073/pnas.93.26.15036

44. Pattison DI, Davies MJ (2001) Absolute rate constants for the reaction of hypochlorous acid with protein side chains and peptide bonds. Chem Res Toxicol 14(10):1453–1464

45. Olson KR, Gao Y, DeLeon ER et al (2017) Catalase as a sulfide-sulfur oxido-reductase: an ancient (and modern?) regulator of reactive sulfur species (RSS). Redox Biol 12:325–339. https://doi.org/10.1016/j.redox.2017.02.021

46. Boyum A (1968) Isolation of mononuclear cells and granulocytes from human blood. Isolation of mononuclear cells by one centrifugation, and of granulocytes by combining

centrifugation and sedimentation at 1 g. Scand J Clin Lab Invest Suppl 97:77–89

47. Quinn MT, DeLeo FR, Bokoch GM (2007) Neutrophil methods and protocols. Preface. Methods Mol Biol 412:vii–viii

48. Kaplan SS, Basford RE, Jeong MH et al (1994) Mechanisms of biomaterial-induced superoxide release by neutrophils. J Biomed Mater Res 28(3):377–386. https://doi.org/10.1002/jbm.820280313

49. Chang S, Popowich Y, Greco RS et al (2003) Neutrophil survival on biomaterials is determined by surface topography. J Vasc Surg 37 (5):1082–1090. https://doi.org/10.1067/mva.2003.160

50. Dooley DC, Simpson JF, Meryman HT (1982) Isolation of large numbers of fully viable human-neutrophils—a preparative technique using percoll density gradient centrifugation. Exp Hematol 10(7):591–599

51. Sivak A, Van Duuren BL (1971) Cellular interactions of phorbol myristate acetate in tumor promotion. Chem Biol Interact 3(6):401–411

52. Van Duuren BL, Sivak A, Segal A et al (1973) Dose-response studies with a pure tumor-promoting agent, phorbol myristate acetate. Cancer Res 33(9):2166–2172

53. Nagy P, Kettle AJ, Winterbourn CC (2010) Neutrophil-mediated oxidation of enkephalins via myeloperoxidase-dependent addition of superoxide. Free Radic Biol Med 49 (5):792–799. https://doi.org/10.1016/j.freeradbiomed.2010.05.033

54. Nagy P, Kettle AJ, Winterbourn CC (2009) Superoxide-mediated formation of tyrosine hydroperoxides and methionine sulfoxide in peptides through radical addition and intramolecular oxygen transfer. J Biol Chem 284 (22):14723–14733. https://doi.org/10.1074/jbc.M809396200

55. Kettle AJ, Winterbourn CC (1989) Influence of superoxide on myeloperoxidase kinetics measured with a hydrogen peroxide electrode. Biochem J 263(3):823–828

56. Kettle AJ, Anderson RF, Hampton MB et al (2007) Reactions of superoxide with myeloperoxidase. Biochemistry 46(16):4888–4897. https://doi.org/10.1021/bi602587k

57. Gallin JI, Seligmann BE (1984) Neutrophil chemoattractant fMet-Leu-Phe receptor expression and ionic events following activation. Contemp Top Immunobiol 14:83–108

58. Timm M, Saaby L, Moesby L et al (2013) Considerations regarding use of solvents in in vitro cell based assays. Cytotechnology 65 (5):887–894. https://doi.org/10.1007/s10616-012-9530-6

Vascular Myography to Examine Functional Responses of Isolated Blood Vessels

Joanne Hart

Abstract

Vascular myography is an in vitro technique used to examine functional responses of isolated blood vessels. This classical pharmacological technique has been in use for over a century. The assay technique studies changes in isometric tone of large and small vessels, arteries and veins, and tissues from genetic or disease models. This chapter describes the apparatus required, tissue collection methods, and the mounting of the tissues in the chambers of both large organ baths and the small vessel myograph. Considerations of the experimental conditions and design are discussed as well as the analysis of the collected data.

Key words Myography, Blood vessels, Endothelial function, Nitric oxide bioavailability

1 Introduction

Modern biomedical science has techniques that can study all aspects of vessel function from single cells to isolated vessels and the vascular bed or in the intact organism. Many experimental methods are available to specifically study blood vessel function. Isolated tissue bath assays are a classical pharmacological tool for evaluating concentration-response relationships. The basic technique has been in use for over 100 years, when it was discovered that isolated tissues and organs remained functional for several hours if kept in oxygenated physiological solutions at physiological temperatures. This assay technique is the basis for the experimental method of determining functional responses of blood vessels, vascular myography, that is described in this article. The method was originally used to study large, conduit vessels (up to 5 mm diameter) but then was later adapted and developed for small resistance-like vessels (100–400 µm) by Mulvany and Halpern [1].

Vascular myography can be used to examine isolated vessels with internal diameters up to about 5 mm, independent of blood flow or nervous system controls. It enables the study of a variety of blood vessels from different species, vascular beds, and pathological

Jerzy Bełtowski (ed.), *Vascular Effects of Hydrogen Sulfide: Methods and Protocols*, Methods in Molecular Biology, vol. 2007, https://doi.org/10.1007/978-1-4939-9528-8_15, © Springer Science+Business Media, LLC, part of Springer Nature 2019

states, which include genetic and disease models. Both arteries and veins may be studied and the assays are robust and reproducible. The technique can provide information on both vascular smooth muscle function and endothelial cell function. Often several organ baths or small vessel myographs are used in parallel to allow simultaneous study of several treatment groups. The method is described here in detail, along with a discussion of considerations for experimental design and data analysis.

2   Materials

2.1   Apparatus and Equipment Preparation (Fig. 1)

1. Essentially the apparatus for organ bath and small vessel myograph experiments are the same. They consist of an organ chamber that is filled with physiological salt solution (usually Krebs solution). The chamber is heated to 37 °C and supplied with 95%O_2 in CO_2 (carbogen) gas. The solution in the bath can be readily replaced by either draining it or using a vacuum mechanism followed by refilling of pre-warmed solution. The tissue is mounted on wires that connect to a micrometer on one side and a force transducer on the other. Changes in force are recorded to a data acquisition system, such as via a PowerLab™ using the LabChart™ software (ADInstruments, www.adinstruments.com).

2. Prior to the experiment, the apparatus is pre-warmed to 37 °C, and force transducers must be calibrated. Calibration should be carried out at the experimental temperate as these transducers tend to be temperature sensitive. Reservoirs containing the

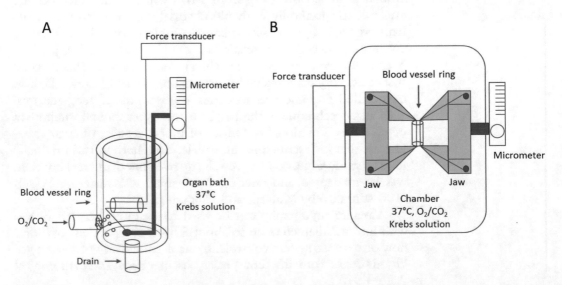

Fig. 1 Organ bath and myograph apparatus. Diagrams of (**A**) organ bath apparatus and (**B**) myograph apparatus

Krebs solution heated to 37 °C and supplied with 95%O_2 in CO_2 (carbogen) gas should be prepared and easily accessed as the chamber solutions needs to be changed often during the experiment.

2.2 Solutions

1. Prior to the experiment, physiological salt solutions, usually Krebs solution and a high potassium-depolarizing solution (KPSS), need to be prepared. Stock solutions of $10\times$ Krebs (without glucose, $NaHCO_3$, or $CaCl_2$), KPSS (without $NaHCO_3$ or $CaCl_2$), $NaHCO_3$, and $CaCl_2$ should be made in advance and stored at 4 °C. The working solutions are then assembled from the stock solutions, with the glucose added on the day of the experiment.

2. Final composition of Krebs solution is (mM) NaCl 119, KCl 4.7, $MgSO_4$ 1.17, $NaHCO_3$ 25, KH_2PO_4 1.18, $CaCl_2$ 2.5, glucose 11.1, EDTA 0.026, and pH 7.4. Ethylenediaminetetraacetic acid (EDTA) should be added to Krebs solution where responses to H_2S are being measured as it chelates metal ions that can facilitate the metabolism of H_2S. The final composition of KPSS is (mM) KCl 123, $MgSO_4$ 1.17, KH_2PO_4 2.25, $CaCl_2$ 2.5, and glucose 11.1.

3 Methods

Once the apparatus is prepared and the solutions made, tissues can be collected, dissected free of fat and connective tissue, and then mounted in the organ bath or myograph (Figs. 1 & 2).

3.1 Tissue Collection and Preparation

These instructions are specifically for collecting a rat aorta, but the same applies for collection of mouse or guinea pig aorta.

1. Cull the rat as approved by the appropriate animal ethics committee. Typically, CO_2 sedation followed by cervical dislocation is used.

2. Perform a complete midline incision, and remove all intestines, stomach, and lungs. Clear away the diaphragm and any other connective tissue. Use swabs to gently blunt dissect any adipose tissue to expose the aorta, from the heart down to the iliac bifurcation.

3. Carefully collect the aorta from the aortic arch down to the iliac bifurcation.

4. Place immediately in ice-cold Krebs solution.

5. Pin out the aorta in a silicone-based petri dish, and keep covered in Krebs solution and on ice.

6. Using fine scissors and forceps, carefully remove all fat and connective tissue from the aorta.

7. Cut the cleaned aorta into 2–4 mm rings using a scalpel blade, ready for mounting in the organ bath.

8. Other vessels, e.g., renal artery, carotid artery, mesenteric artery, and cerebral arteries, may also be collected; the important points are to dissect carefully and keep the tissues in ice-cold Krebs as far as possible. Once collected the tissue can be pinned out in a silicone-based petri dish, in ice-cold Krebs solution. Small arteries can then be cleared of connective tissue and adipose tissue under a dissecting microscope (40×).

3.2 Mounting the Vessels in the Organ Bath (Fig. 1A & 2)

1. Carefully thread the vessel onto the bottom hook, taking care not to scrape the inside (endothelium) or cause damage to the vascular smooth muscle.

2. Thread through the wire to the force transducer, taking the same due care. Do not cross the wires; they should be parallel. Immerse in the organ bath, and set the resting tone (*see* Fig. 2 and Subheading 3.4).

3.3 Mounting of the Vessels in the Myograph (Fig. 3)

1. Segments of the artery 2 mm long are cut with small sharp Vannas scissors. A piece of wire (2 cm, 50 µM diameter) is threaded through the lumen.

2. In the myograph bath, in ice-cold Krebs solution, the ends of the wire are fixed to the screws on one side of the myograph jaws, which are open, positioning the vessel in the groove.

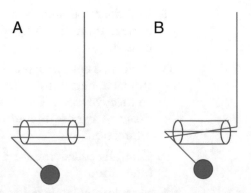

Fig. 2 Correct mounting of vessels in the organ bath. Isolated blood vessel ring mounted on the fixed wire at the bottom and a wire connected to a force transducer at the top. (**a**) Correctly mounted ring with wires parallel through the vessel lumen, (**b**) incorrectly mounted ring with wires crossed through the vessel lumen

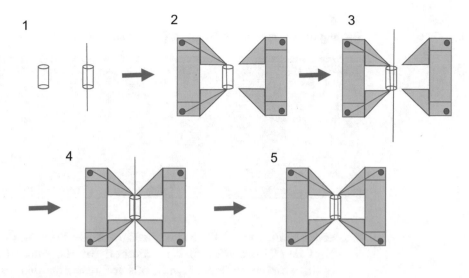

Fig. 3 Method for mounting vessels in the myograph. Diagram showing the mounting of small vessel rings in the myograph. (1) A wire is threaded through the lumen of the blood vessel ring. (2) This wire is connected to the screws on the myograph jaw. (3) A second wire is threaded through the lumen. (4) Myograph jaws are closed to trap the second wire. (5) The second wire is fixed to the second jaw, and then the myograph jaws are separated, so the vessel is fixed with no tone (0 mN)

3. Once the first wire is in place, the second wire is very carefully threaded through the lumen, taking care not to scrape the inside of the vessel. It is best done with one smooth movement. Do not cross the wires.

4. The jaws are closed, which hold the second wire in place, while it is fastened under the screws at the front and rear.

5. The jaws are then opened to ensure the vessel is fastened, but there is no tone applied. The myograph bath is then warmed to 37 °C and supplied with carbogen.

3.4 Setting the Resting Tone

1. The initial resting tone can influence the reactivity of the vessel, so it is important to standardize initial conditions to ensure the vessels being assessed are done so in a reliable and consistent manner.

2. For large conduit arteries, a length-tension curve can be constructed to assess the contractile capacity of the vessel. From this data, an exponential curve can be fitted and apparent transmural pressure calculated. However, for larger arteries, it is usual to use previously calculated and published optimal resting tone (*see* Table 1).

3. For smaller vessels (100–400 μm), a length-tension curve (normalization protocol) is constructed in each vessel. From this curve, the effective transmural pressure and vessel diameter (D_{100}) are calculated [1]. If using the LabChart™ data

Table 1
Resting tone values for different large or conduit vessels

Tissue	Resting tone	References
Rat aorta	1 g	[2]
Guinea pig aorta	1.2 g	[3]
Rabbit aorta	2 g	[4]
Pig coronary artery	5 g	[5]
Human coronary artery	6–8 g	[6]

acquisition, the DMT Normalization Add-On™ module for LabChart™ provides an easy method for the calculation of optimal precontraction conditions for resistance-like arteries or any small tubular tissue research. The vessels are each set at a tension equivalent to that generated at 0.9 times the D_{100} at 100 mmHg and left to equilibrate for 20 min.

3.5 Standard Experimental Protocols (Fig. 4)

1. A standard experimental protocol for determining a concentration-response relationship would be to firstly set resting tone and then wait for 15 min.

2. Assess vessel viability, by obtaining a tissue maximum contractile response, and wash.

3. Obtain endothelial function response, and wash (optional).

4. Add inhibitors, and wait for the required time (usually 30 min).

5. Carry out the concentration-response curve by adding the agonist in cumulative 0.5 log concentration increments. For assessing vasorelaxation responses, the vessels need to be precontracted to about 50% of their maximal contraction, prior to adding the vasorelaxant. At the end of the CRC, it is usual to obtain a maximal vasorelaxation response using a NO donor (e.g., sodium nitroprusside, SNP 10 μM), a K^+ channel opener (e.g., levcromakalim, LKM 10 μM), or a Ca^{2+} antagonist (e.g., nifedipine, 10 μM). This value is used to normalize vasorelaxation responses across groups. (*see* **Notes 1–3**).

6. Finally, obtain another tissue maximum response. This is optional but will ensure that there has been no loss of vessel viability over the course of the experiment.

3.6 Assessment of Vessel Viability

1. Allow the vessel to equilibrate at basal tone for about 20 min.

2. Test the vessel contractility with KPSS or an α-adrenoceptor agonist (e.g., phenylephrine, 10 μM) or thromboxane receptor agonist (U46619, 1 μM). The latter is preferred for mouse vessels, as this tissue does not consistently respond to the KPSS.

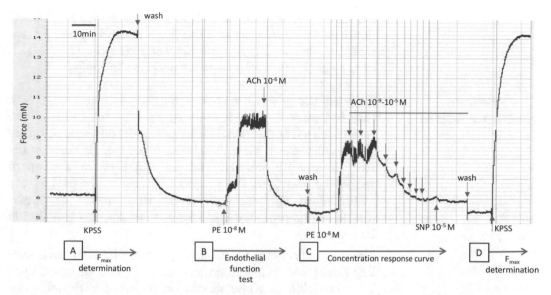

Fig. 4 Standard experimental protocol for a vasorelaxation response effect. Reproduced laboratory trace showing a ring of rat carotid artery, which has undergone the normalisation procedure (*see* Subheading 3.4, **step 3**). (**A**) The experiment commences with the determination of the maximal contractile activity of the tissue (F_{max}), using KPSS. This is washed out, and then (**B**) the endothelial function assay is performed. Phenylephrine (PE) is added to precontract the vessel, and then a single concentration of ACh is added. The next step is (**C**) the concentration-response curve; here PE is again added to precontract the vessel and then ACh added in increasing concentration. SNP is added to elicit a full relaxation (R_{max}). Finally, (**D**) the tissue contractility is retested with KPSS to ensure no change in tissue viability throughout the experiment

3. The vessel should significantly contract. (*See* **Note 4**).

4. Wash out the contractile agent and replace with Krebs.

3.7 Concentration-Response Curves

1. Most organ bath vascular myography experiments involve comparisons of responses to agonists in the presence or absence of an antagonist. This is usually done by constructing a cumulative concentration-response curve (CRC). Large concentration ranges are required so drug concentrations are usually presented on a logarithmic scale. This typically produces a curve of sigmoidal shape, with the 20–80% response being linear.

2. For vasoconstrictor CRC, start at a low concentration, and add the next concentration once the previous contraction has peaked. Be aware of the potential for desensitization (tachyphylaxis) that may occur with some agonists (particularly peptides like angiotensin II).

3. For vasodilator curves the vessel is first contracted to approx. 50% of the maximum, using a suitable agonist, usually an α-adrenoceptor agonist (e.g., phenylephrine). The vasodilator is then added in a cumulative manner, adding the next dose at the nadir of the previous dose. Usually the response to a

standard concentration of a known vasodilator (e.g., the NO donor sodium nitroprusside, the K_{ATP} channel opener levcromakalim, or the L-type Ca^{2+} blocker nifedipine (10 μM)) is used to elicit 100% relaxation for purposes of analysis.

3.8   Assessment of Endothelial Function

1. To assess the role of the endothelium in vascular responses, it may be removed. This is done by rubbing the inside of the vessel with a wooden stick (for large vessels) or a wire or suture (for small vessels). Care should be taken not to damage the vascular smooth muscle, and an endothelium assessment should be added to the protocol to ensure removal of the endothelium. Alternatively, vessels from models of disease may have compromised endothelial function, and this should be assessed.

2. To assess endothelial function, contract to 50% KPSS max with PE. Add 1 μM ACh and endothelium-dependent vasodilator. The relaxation response should be abolished with deliberate endothelium removal (indeed a contraction response may be seen). A relaxation response of >80% indicates the presence of a functional endothelium that has not been damaged by the dissection or mounting process. A lesser endothelium-dependent relaxation response may be expected in vessels from models of cardiovascular disease.

3.9   Assessment of NO Bioavailability (Fig. 5)

1. To assess the ability of the endothelium to produce NO, the NO bioavailability assay is used [7, 8]. Vessels are precontracted to 20–30% max using an α-adrenoceptor agonist (e.g., phenylephrine), and then an eNOS inhibitor (e.g., N-nitro-L-arginine methyl ester (L-NAME) 100 μM) is added. The ensuing contractile response is indicative of the activity of NO that is being released from the vessel.

3.10   Assessing Effects of Superoxide Anion Production on Vascular Responses

1. To assess the acute effects of superoxide anions ($O_2^{\cdot-}$) on vascular function, $O_2^{\cdot-}$ can be generated in the chamber by using either pyrogallol (100 μM) [2, 9], which autoxidizes and releases $O_2^{\cdot-}$, or the combination of hypoxanthine (100 μM) and xanthine oxidase (0.1 U/ml) [9].

2. Superoxide dismutase (SOD, 250 U/ml) is used as a control.

3. The $O_2^{\cdot-}$ generating treatment needs to be added immediately prior to the assessment of the functional response of interest.

3.11   Examining Responses to Hydrogen Sulfide and H₂S Donors

1. Hydrogen sulfide is a gas at room temperature and pressure and needs to be dissolved in H_2O or physiological solution prior to use. It is highly volatile, and adding it to physiological solution in organ baths is a problem as the concentration declines rapidly at 37 °C in the presence of O_2 (Fig. 6).

2. H₂S-releasing compounds are commonly used. There are broadly two types of compounds, the sulfide salts and the

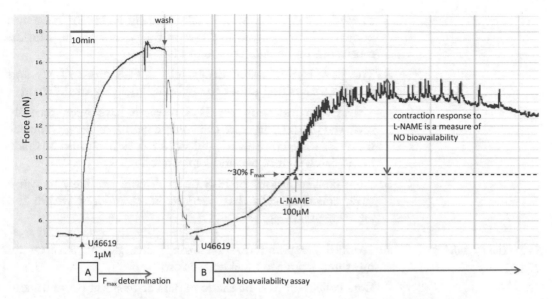

Fig. 5 Experimental protocol for determining NO bioavailability. Reproduced laboratory trace showing a ring of mouse aorta, which was set at 5 mN basal tone. (**a**) The experiment commences with the determination of the maximal contractile activity of the tissue (F_{max}), using the thromboxane receptor agonist U46619. This is washed out, and then (**b**) U46619 is added to precontract the vessel to 30% F_{max}, followed by a single concentration of L-NAME (100 μM). The ensuing contraction is used as a measure of NO bioavailability [7, 8]

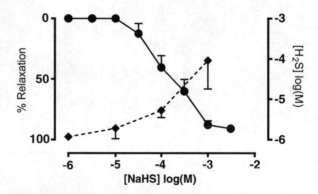

Fig. 6 Vasorelaxation response and bath [H_2S] after delivery with NaHS. Cumulative concentration-response curve to H_2S donor NaHS in mouse aortic rings (•, $n = 7$) precontracted with the thromboxane mimetic U46619 to elicit 50% F_{max} precontraction. Responses are expressed as percentage reversal of the level of precontraction (left y-axis). Right y-axis shows concentration of H_2S detected in the organ bath from cumulative additions of NaHS (1 μM^{-1} mM, $n = 5$) determined using an H_2S-sensitive probe (ISO-H_2S, WPI) connected to the Apollo 4000 Free Radical Analyzer (WPI). The probe was calibrated using a solution of Na_2S, under anoxic conditions in a closed chamber, at 37 °C immediately before use. The concentration of H_2S delivered from increasing concentrations of NaHS was then measured in oxygenated Krebs solution at 37 °C [12]. All data are given as mean ± sem

214   Joanne Hart

slow-release donors. Sulfide salts, usually NaHS or Na$_2$S are known as "fast" releasers of H$_2$S as they dissociate in solution causing an immediate release of H$_2$S.

3. The slow H$_2$S-releasing drugs release H$_2$S more slowly and provide a sustained rather than "bolus" delivery of the drug.

4. Pharmacological tools for H$_2$S research have been recently reviewed [10]. In all cases the volatile nature of the gas mediator needs to be considered, and stock solutions should be made immediately before use and kept on ice.

5. High concentrations of H$_2$S or its donors are likely to cause toxic effects in cells that need to be considered within the experimental design.

3.12 Data Analysis (Fig. 7)

1. From the experimental recording, changes in tension are measured from the baseline tension.

2. Data collection may be made in data acquisition program and then exported to an Excel spreadsheet for manipulation.

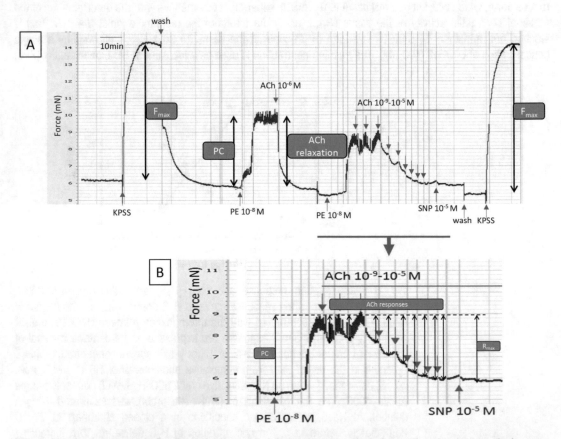

Fig. 7 Standard data analysis. Reproduction of a laboratory trace that shows the standard analysis parameters. (**a**) F_{max} calculation (baseline to top of the KPSS response), endothelial function assessment (ACh relaxation/PC) $\times$ 100, and the final F_{max} calculation (baseline to top KPSS response). (**b**) Shows an expansion of the concentration-response curve, where the vasorelaxation response to ACh for each concentration is calculated as (response/(PC $-$ R_{max})) $\times$ 100, as the responses are normalized to the R_{max} determination

3. Contraction responses are normalized to the tissue maximum (F_{max}), which was obtained at the commencement of the experiment (*see* Subheading 3.5, **step 2**).

4. Relaxation responses are calculated as % relaxation, normalized to the response of a known vasodilator ($R_{max} = 100\%$).

5. The precontraction value for the vasodilator curve is calculated as a % of the tissue maximum and should be 40–60%.

6. Occasionally, data collection may be complicated by spontaneous tone oscillations. This can be managed by taking the mean response at each data point or the top or bottom of the oscillation, keeping this consistent throughout the data analysis. A slow decline in precontracted vessel tone or "fade" may also be observed; in this case some discretion should be used to separate fade from vessel responses.

7. The F_{max} and % precontraction data are important experimental parameters that should not differ between treatment groups. Thus, a table of these values for each treatment group should be provided as supplementary data for all published work (Table 2). This table should also include n values and may include the endothelial function assessment data. A statistical analysis (*t*-test or one-way ANOVA) should be carried out to ensure adequate matching between the groups.

3.13 Curve Fitting

1. Once the concentration-response curve data has been collected, it is normal to fit a sigmoidal logistic curve to the data to determine the EC_{50} response. This is the concentration at which 50% of the total response has occurred. This analysis can be easily carried out in GraphPad Prism™ or another program via nonlinear regression. This is the standard procedure for where there is an agonist-receptor relationship, so this may

Table 2
Example data table for collecting experimental parameter data

Vessel	n	Tissue maximum contraction (F_{max})	Precontraction level (%PC)	Endothelial function check (% ACh (10 μM) response)
Control				
Treatment 1				
Treatment 2				
p Value		ns	ns	

not be appropriate for agents without a known receptor, e.g., H_2S. In this case, a two-way ANOVA may be used to compare the concentration-response curves.

4 Notes

1. More than one concentration-response curve may be carried out in a daily vascular myography protocol; however appropriate time control experiments must be carried out.

2. Some compounds are irreversible and cannot be washed out. This must be accounted for in the experimental design.

3. Do not over-contract vessels prior to construction of vasodilator concentration-response curves. There is good evidence that in tissues contracted over 50% F_{max} functional antagonism occurs and this will invalidate the experiment [11]. Aim for 50% of the F_{max} for precontraction.

4. A strict set of inclusion criteria should be set for each experiment series. This will include minimum acceptable tissue max responses, minimum acceptable endothelium responses, and range of acceptable precontraction responses. For example, rat aortic rings appropriate inclusion criteria could be tissue max >1 g, endothelial function >80%, and % precontraction 40–60%. These values should be set during the experiment design, and the data are presented (*see* Subheading 3.6) in the results or supplementary results sections of publications.

Acknowledgments

Reproduced traces shown are adapted from original LabChart recordings generated by Ms. Suzan Yildiz and Mr. Jafer Al Qaeisoom.

References

1. Mulvany MJ, Halpern W (1977) Contractile properties of small arterial resistance vessels in spontaneously hypertensive and normotensive rats. Circ Res 41(1):19–26

2. Leo CH et al (2012) Endothelium-dependent nitroxyl-mediated relaxation is resistant to superoxide anion scavenging and preserved in diabetic rat aorta. Pharmacol Res 66 (5):383–391

3. Jones RL, Woodward DF (2011) Interaction of prostanoid EP(3) and TP receptors in guinea-pig isolated aorta: contractile self-synergism of 11-deoxy-16,16-dimethyl PGE(2). Br J Pharmacol 162(2):521–531

4. Hart JL, Sobey CG, Woodman OL (1995) Cholesterol feeding enhances vasoconstrictor effects of products from rabbit polymorphonuclear leukocytes. Am J Phys 269(1 Pt 2): H1–H6

5. McPherson GA et al (1999) Functional and electrophysiological effects of a novel imidazoline-based K(ATP) channel blocker, IMID-4F. Br J Pharmacol 128(8):1636–1642

6. Berkenboom G et al (1989) Comparison of responses to acetylcholine and serotonin on isolated canine and human coronary arteries. Cardiovasc Res 23(9):780–787

7. Leo CH, Hart JL, Woodman OL (2011) Impairment of both nitric oxide-mediated and EDHF-type relaxation in small mesenteric arteries from rats with streptozotocin-induced diabetes. Br J Pharmacol 162(2):365–377

8. Leo CH, Hart JL, Woodman OL (2011) 3',4'-Dihydroxyflavonol reduces superoxide and improves nitric oxide function in diabetic rat mesenteric arteries. PLoS One 6(6):e20813

9. Al-Magableh MR et al (2014) Hydrogen sulfide protects endothelial nitric oxide function under conditions of acute oxidative stress in vitro. Naunyn Schmiedeberg's Arch Pharmacol 387(1):67–74

10. Papapetropoulos A, Whiteman M, Cirino G (2015) Pharmacological tools for hydrogen sulphide research: a brief, introductory guide for beginners. Br J Pharmacol 172(6):1633–1637

11. Stork AP, Cocks TM (1994) Pharmacological reactivity of human epicardial coronary arteries: characterization of relaxation responses to endothelium-derived relaxing factor. Br J Pharmacol 113(4):1099–1104

12. Al-Magableh MR, Hart JL (2011) Mechanism of vasorelaxation and role of endogenous hydrogen sulfide production in mouse aorta. Naunyn Schmiedeberg's Arch Pharmacol 383(4):403–413

INDEX

A

Acetylcholine (Ach)..................92, 94, 98, 101, 103, 131
Adenylyl cyclase.. 142, 143
Aging 10, 140, 144, 197
Albumin.. 45, 52, 54, 58,
    60, 61, 65, 70, 90, 92, 94
Anesthesia 111, 113, 175, 176
Angiogenesis...v, 2, 6, 20,
    151, 152, 158, 164
Aorta 89, 91, 92, 94,
    95, 100–104, 134, 138, 141, 174–176, 207, 208,
    210, 213
Arachidonic acid 144
Arterial pulse waveform (APW) 109–122
Artery
    carotid 20, 27, 30, 33,
        110, 112, 114, 118–123, 208, 211
    mesenteric..20, 27, 30,
        33, 89, 91, 98, 100, 101, 105, 125, 127, 129, 208
    uterine.. 20, 27, 30, 33
Asolectin 93, 94, 96, 102
Atherosclerosis ....v, 2, 180
ATP-sensitive potassium channels (KATP) 4,
    20, 79, 82, 143, 145, 212
Autonomic nervous system 167

B

Bacteria 3, 138, 173
Biotin pulldown assay 52
Biotin-Switch method...38, 41
Blood pressure... 90, 91,
    109, 111, 113, 114, 117, 120, 121, 174

C

Carbon monoxide (CO)......................1, 19, 37, 137, 138
Catheterization 175, 176
Caudal ventrolateral medulla..167
Central nervous system 138
Chemoinvasion.. 155, 160, 161
Chlorinating activity 182, 184, 187, 190, 191
Circumventricular organs 167, 168
Colon 138, 173, 175, 177
Coomassie dye...60, 64
Corpus cavernosum 137–146

D

Cyclic adenosine monophosphate (cAMP) 142–143
Cyclic guanosine monophosphate (cGMP)........ 141–144
Cyclooxygenase (COX) 144
Cystathionine (Cysta) 2, 3, 10, 11
Cystathionine β-synthase (CBS)......................2, 37, 152
Cystathionine γ-lyase (CSE)......................................2, 19
Cysteine 2–4, 10, 25,
    38, 42, 45, 46, 58, 70, 72, 90, 137, 138
Cysteine aminotransferase (CAT) 3, 19, 140

D

D,L-propargylglycine (PAG).............................. 152, 158,
    159, 161, 164
Dephosphorylation 39

E

Electrode 99, 100,
    127, 131, 135, 168, 173, 175, 187
Electrophoresis.............................. 28, 65, 156, 162, 163
Endothelial cells 6, 34, 141,
    153, 157, 158, 160, 163, 164
Erectile dysfunction ....v, 140, 142, 144
Erectile function 137, 138, 140–142, 144, 145
Estrogen 20, 21

F

Fluorescent dye 79–81, 87, 154, 165

G

Gasotransmitter.......................... v, 1, 4, 6, 9, 19, 37, 125
Gelatin 24, 30, 80,
    82, 85, 86, 153, 157, 161–164
Glutathione (GSH) 2–5, 41, 53, 69, 90
Granulocytes.................... 180, 183, 184, 192, 195, 199
Gut 3, 173
GYY4137 ....20, 164

H

Halogenation...................................180, 181, 189, 198
Heart.................................. v, 5, 20, 91, 111, 138, 207
Hemodialysis 10
Homocysteine (Hcy) 2, 3, 10, 11, 13, 15
Homolanthionine (Hla) 2, 10–13, 15
Homoserine (Hse) 11–13, 15

Jerzy Bełtowski (ed.), *Vascular Effects of Hydrogen Sulfide: Methods and Protocols*, Methods in Molecular Biology, vol. 2007,
https://doi.org/10.1007/978-1-4939-9528-8, © Springer Science+Business Media, LLC, part of Springer Nature 2019

Hyperpolarization 79–81, 83, 84
Hypertension .... v, 2, 90, 140, 144, 152

I

Immunoblotting 21, 24, 29, 30, 65
Immunofluorescence 21, 30–32
Inflammation 180
Intracavernosal pressure 138, 140, 143–145
Intracerebroventricular 167, 168, 171
Intracolonic administration 173, 176
Iodoacetic acid (IAA) 43

J

Jugular vein 111–113, 117–119

K

Krebs solution 91, 92, 94–99, 145, 206–208

L

Lanthionine 2, 9–16

M

Maleimide 41, 42
Mass spectrometry 9–16
Matrigel 156, 163–165
Menstrual cycle 20
Methylene blue assay 21, 32, 33
Microorganisms 180
Microscope 25, 85, 86,
     127, 129–131, 152, 154–157, 160, 164, 208
Migration 151, 154, 160, 165
Minipumps 168, 170
Multiple reaction monitoring 9–16
Myeloperoxidase (MPO) 5, 180–200
Myograph 89, 100,
     125, 127, 129, 131, 133–135, 206–209
Myography 20, 205, 211, 216

N

Neutrophil 5, 180,
     182–184, 192, 193, 195, 198, 200
Nitric oxide (NO) 1, 19,
     37, 39, 90, 101, 137, 138, 142, 145, 151
Nitrosopersulfide (SSNO-) 5
Nitrosothiols 5, 38, 43, 90, 94, 96, 102
Noradrenaline 91, 94–99, 104
Nucleophilicity 41

P

Paraventricular nucleus 167, 171
Parkin 4

Penile erection 137, 138, 140, 141
Permeability .... v, 6, 151
Peroxidase cycle 181–183, 186, 195
Peroxynitrite 5, 38
Persulfidation 4, 38, 39,
     42, 43, 45, 46, 52, 56, 58, 60, 61, 65, 72
Persulfides 3–5, 39–46,
     52, 53, 60–62, 64–67, 69–71, 73
Phagocytosis 180
Phenylephrine (PE) 92, 94,
     97, 103, 123, 145, 210–212
Phosphodiesterase (PDE) 142, 143
Phosphorylation 39
Polysulfides (H$_2$S$_n$) 5, 52, 58, 100, 109, 115
Pregnancy 20, 21
Pressure transducer 110, 112, 113, 115, 120
Proliferation 151, 154, 159–162, 165
Proteomics 43, 45, 55, 63–66, 72, 74
Pulse wave velocity (PWV) 114, 122
Pyridoxal 5'-phosphate 2, 25, 32

R

Real-time quantitative PCR (qPCR) 21
Reperfusion 5, 180
Rostral ventrolateral medulla 167

S

Sildenafil 142, 144
Smooth muscle cells .... v, 2, 6, 79–87
S-nitrosation 38, 39
S-nitrosoglutathione (GSNO) 5, 92–96, 110, 111
Sodium dodecyl sulphate-polyacrylamide gel
     electrophoresis (SDS-PAGE) 23, 28,
     47, 60, 62, 64, 66, 156, 163
Sodium hydrosulfide (NaHS) 25, 33
Sodium sulfide 58, 110
Soluble guanylate cyclase 5, 95, 96
Spontaneously hypertensive rat (SHR) 90, 103
Staining 54, 55,
     61–66, 71, 73, 74, 154, 155, 193
Stereotactic coordinates 168
Sulfane 39, 56, 58
Sulfenic acid (-SOH) 5
Sulfide:quinone oxidoreductase (SQR) 4
Sulfur 3, 4, 10, 37, 39, 52, 56–58, 127
Sulfuration 4
Superoxide 5, 39, 185, 200, 212

T

Tag-switch method 42, 45
Thiol 2, 3, 19, 38, 41–44, 90, 94
Thionitrous acid (HSNO) 5
Thioredoxin (Trx) 3, 5, 39, 45, 53, 69

Thioredoxin reductase (TrxR) 39, 45, 69
3-Mercaptopyruvate sulfurtransferase
(3-MST)... 3, 19, 138, 142

U

Ubiquitin 4, 38
Uremic toxins..9–16

V

Vasoconstriction .... v, 5, 89, 92, 99, 144

Vasodilation 21, 143, 151
Vasorelaxation ....v, 5, 89,
92, 98, 101–104, 210, 211, 214

W

Wistar rat 90
WSP-1 .... 80–82, 85, 86, 154, 158, 159, 165

Z

Zymography 155, 156, 161–163

Printed in the United States
By Bookmasters